DIGITAL MARKETING TECHNOLOGY IN AUTOMOTIVE INDUSTRY – by HIMANSHU D SHAH

Executive Office:
219 Changebridge Road
Montville, New Jersey 07045

Technology Office:
510 Thornall St., STE 320
Edison, New Jersey 08837

Table of Contents

Cover Sheet /Table of Contents………………………………………………………………….….2

Prologue……………………………………………………………………………………………….4

Executive Summary – Business Overview……………………………………………………...6

Success Factors……………………………………………………………………………..…..…6

Introducing eznewcar.com……………………………………………………………………….7

The Competition……………………………………………………………………………………8

So, How Does it Work?..10

How Did This Come to Be?..16

Milestones…………………………………………………………………………………………37

How is it Going to Grow?...38

Pricing Models for the Initial Newspapers……………………………………………………39

Known Elements and Assumptions…………………………………………………………...43

Build it and they will come…… Not (EZNEWTRUCK)..44

Invest to Date……………………………………………………………………………………..45

Social Media & Chief Technology Officer Bio………………………………………………..46

Dealership Back End Functionality……………………………………………………………47

One Last Point in Support of eznewcar.com…………………………………………………47

Dealer Manual…………………………………………………………………………………….48

Sales Manual……………………………………………………………………………………..72

Eznewcar.com

Sales, Customer Service   IT Technology

219 Changebridge Road   510 Thornall Street Suite 320

Montville, NJ 07045   Edison, NJ 08837

Telephone: 201 602 0628

Fax: 973 263 0832

E mail: himanshu.shah@eznewcar.com /
hshah@hpinfosystem.com //*support@hpinfosystem.com*

No offering is made or intended by this document. Any offering of interests in eznewcar.com will be made only in compliance with Federal and State securities laws.

This document includes confidential and proprietary information of and regarding eznewcar.com. This document is provided for informational purposes only. You may not use this document except for informational purposes, and you may not reproduce this document in whole or in part, or divulge any of its contents without the prior written consent of eznewcar.com. By accepting this document, you agree to be bound by these restrictions and limitations.

Prologue:

I know there are countless ways to build what the experts consider a winning business plan, I also know that investors have the opportunity to review many business plans full of hope and excitement that they will have the proper arrangement of words so to convince someone to have faith in their vision and make an investment. I realize that in order for a plan to be taken seriously, it must fit into one of these accepted styles of formats. I know this, but I think the eznewcar.com story needs to be told and sold in a different format. To fully understand the significances of what this business is, and will become, you need to understand where it came from and why it was built. This is a website that has its origins in time tested methods that go back to the very beginning of retail automobile marketing itself. But today, that industry is under assault from the Internet. The Internet and the digital marketing phenomenon have changed shopper's habits and buying processes forever. The traditional marketing industry has felt the effects, and the agencies that once thrived on the robust marketing venue of the print world have now been forced to look to other places to supplement the lost income. The idea that produced eznewcar.com began in the ashes of the print marketing industry and will rise from there like a phoenix. And from these ashes, eznewcar.com will rise to become the catalyst that allows for a change in attitudes and processes in a hold-out industry that begs for full transparency.

For some thirty years I have worked in and run a number of retail automobile dealerships in one of the most competitive regions of the United States, the New York and New Jersey Metro Market, and for the past twelve plus years, I have owned an advertising agency that has specialized in retail automotive advertising and marketing. My experience is huge and varied. I worked for small "mom and pop" single point stores, all the way to the pinnacle of Potemkin Cadillac in Manhattan where high-volume retail sales were invented and part of the publicly traded AutoNation retail conglomerate in Flemington Car & Truck Country in New Jersey. I performed the duties of Controller and General Manager working at the very top level of management and operations. As a Chrysler Break-through dealer in the mid 90's, one of only 50 in the US, I interacted with the top of Chryslers Corporate management, traveled to and benchmarked some of the best operated businesses (Ritz Carlton, Harley Davidson, Penske Organization, to name a few), and built "best practices" reports on some of Americas best auto dealer chains and my own operations. In the marketing industry, I have worked hand and hand with the dealers and the radio, cable and newspaper organizations creating marketing plans, creative materials and campaigns for every facet of the dealerships. Over these forty-two years I have seen firsthand how the industry, on many levels has changed from where it was, in the 1970's, to where it is, and where it is going. My vast experience has given me the ability to connect the dots, if you will, and see the future of automobile marketing and sales from the unique perspective of both sides of the table....from within, and from the outside.

It is this experience and vision, along with a group of highly talented programmers, administrators and internet masters that has lead to the creation and development of eznewcar.com. In a number of business and entrepreneur articles and magazines I have read others who say that it is the "idea" that is the easy part, getting it operational and funded, now that is the difficult part. I could not agree more than I do. I look at eznewcar.com as an obvious and extremely logical next step in the evolution of automobile marketing and sales. The challenge is convincing you that I am correct. I believe, and countless others who have seen the facts, that eznewcar.com will have success on a national level and become the dominant player in that arena. Of that, I have no doubt because the concept is so well thought out, and so well developed. Many dream….. You dream, I dream, but the difference is that some of us dream in Technicolor. I see the full potential of eznewcar.com on the national stage, changing the auto retail sales process forever. Just look to Elon Musk, with Tesla Motors. Many bet he could not produce a world class electric powered sedan, but against the odds, he has created a beautiful new automobile that has captured the imagination of many in the auto buying public, and not content to just shake up the world with a new propulsion system in a spectacular looking sedan, he has taken on the franchise system and dealer representative bodies with a different approach to how he wants it marketed to the consumer…..that may or may not be successful, but that is not the point, Tesla Motors is diving into the deep end of the pool convinced it can change perceptions and so far, they are succeeding because they believe in the product, the process and themselves. Hoping for, and wishing success, does not happen by itself, it comes from a well thought out idea, with a true need in the marketplace coupled to a solidly constructed business plan, hard work, dedication and individuals who can have the vision to see into the future and help shape its direction.

Executive Summary

Business Overview

We are...

- A group of automotive and information technology professionals with over 70 years of combined experience who are backed by an alliance of global industry partners in the automotive retail industry, social media, software development and energy sectors.

We have...

- Identified the serious shortcomings of the current online marketing strategies available to automotive dealers and developed the most advanced platform to provide consumers who are ready to purchase an automobile with the precise information they need to make the most informed buying decision possible.

We will...

- Provide dealers and newspapers with the most cost effective and direct channel to reach consumers and communicate their best and most current offers while providing auto buyers an easy and simple one-stop venue to review the best offers from a variety of car dealers, all while completely anonymous, and only then contact the dealer to schedule delivery.

Success Factors

Eznewcar.com is uniquely qualified to succeed due to the following reasons:

- Eznewcar.com is the first and only on-line automobile advertising portal that allows a shopper to view a dealers best advertised price for any given automobile or truck expressed as a purchase price, a monthly finance payment or as a lease payment all while maintaining the shoppers complete anonymity.
- Eznewcar.com allows a customer to shop and compare any advertised vehicle against any other advertised vehicle, regardless of make, model or dealer without having to physically go from dealer to dealer thereby saving time and anxiety.
- Eznewcar.com will be the first on-line automobile advertising portal that will facilitate a true "On Line" automobile buying experience.
- Eznewcar.com is ideally suited for newspaper on-line sales organizations. Newspapers will be able to recoup lost revenues and compete with the other "pure-play" sites with a superior product that is priced thousands below its competition.
- In its ultimate configuration, Eznewcar.com will allow a customer, from the comfort of their living room, compare and select a new vehicle. Once the vehicle is selected, it will then if the purchaser so chooses, run a "soft credit" pull, tier his credit level, adjust the price if necessary and even allow

- the customer to trade in a vehicle with a calculated trade allowance and factor in any pay-off on that trade all while adding the appropriate state, county and local sales tax into the final numbers.
- For the automobile dealer he will have, for the first time, an on-line portal where he can advertise his best deals without being limited by space as in a traditional newspaper advertising space.
- The automobile dealer will have a truly cost effective advertising medium with a one-time monthly cost that is thousands less than he pays for traditional newspaper advertising; eligible for factory co op reimbursement and easy to instantly up-date should the factory offer additional rebates during a month for a special promotion.
- Customers when surveyed overwhelmingly state that they truly dislike the current automobile buying experience and would prefer to eliminate the negotiation process altogether, eznewcar.com is the only site set up to accomplish that with all three types of pricing, purchase, finance and lease.
- Today's savvy internet shopper has been influenced by sites like Amazon.com, Hotels.com and Progressive.com where many competing brands and offers are placed side by side to allow the consumer to evaluate which is best for him, eznewcar.com is the only automobile shopper portal that allows that.
- Eznewcar.com will appeal to the younger and less experienced auto buyer because it will allow them to make an intelligent decision on which is the best deal for them without the fear that they did not actually get the best deal as they see all of the offers available to them.
- Eznewcar.com's web site is completely devoid of annoying, distracting and unrelated pop-up and banner advertising. No dealer gains a better or stronger position on any page than any other dealer. We level the playing field for dealers and consumers alike. Dealers let their best deals do the talking.
- We have shown eznewcar.com to the legal counsel for NJCAR who have given their blessing to the program. NJCAR, who represent all of the dealers, cannot indorse any one product or company, but they have seen the entire program and have given us their "approval" which we think is a major step up for us.

If I haven't lost you yet, let's take a few minutes here to review exactly what eznewcar.com is, how it works and what it promises. From there, we can look at the industry from the perspective of where it was, is and is headed and how eznewcar.com fits into that world. We will review the few miss-steps that a new company invariably makes and how we corrected them. Finally, we will review the accumulated costs to get us where we are, to grow this to its potential and then the kind of revenue streams that are expected and projected.

Introducing eznewcar.com – A Product of HP Infosystem

Eznewcar.com incorporates all of the necessary components to facilitate the accurate marketing of a dealerships inventory, at each vehicles optimum

selling price, while allowing the consumer the ease of shopping on line in complete anonymity to view and compare each vehicle with other vehicles from competing dealers and manufacturers. We have taken the "Amazon.com" shopping model, so to speak, of side by side product and price comparisons and brought it to automobile shopping but then taken it a giant leap beyond that. Eznewcar.com was originally introduced on 2014 when it went live for the first time. It opened up with 29 dealers onboard, but no real marketing in place yet. In 2014 to 2017, we had a presentation booth at the Atlantic City Digital Dealer Conference where we had the opportunity to show our product to dealers basically from east of the Mississippi river for the very first time. The dealers that viewed it were impressed, I am happy to report and many gave us their contact info so we can contact them when eznewcar.com opens in their region. We had plans to go to the September Digital Dealer Expo in Las Vegas in September, but because of the revision to the programming, decided to hold off and participate in the NADA convention in Las Vegas next year, as we will have the full-blown improved system available to show. Besides, the NADA Convention is a better venue because you get the dealers and main decision makers present at that show, along with their checkbooks. The Digital Dealer show tends to attract the Internet Managers and Sales Managers who can see the product and report back to upper management, but are not able to make a purchase decision, so with a limited budget, we will spend our money more wisely.

The Competition….(there really isn't any).

The other 3rd party websites available today:

Currently there are 5 main 3rd party websites in addition to eznewcar.com. Cars.com is the oldest site along with Autotrader.com. These two sites have been the mainstays of the industry for the past 10 years or so. But sadly for them, they have remained static choosing to not improve or innovate their positions. As their product became old and stale, they have faced many defections and have opened the door to newer, more ambitious players in this market segment, eznewcar.com being the newest, and in my humble opinion, the best. Both sites are very similar in that both sweep the dealers inventory and for the most part, take the dealers inventory at MSRP value, display them on their website and allow the consumer to search for a vehicle buy price, brand and model, but remember, they search based upon the posted price by the dealer. Also, both of these sites are used car centric, so they do a very nice job with a used car, but not so much with a new car. It is because each used car is unique, for the most part, no two are exactly the same, different equipment, colors, mileage and condition, so the price is the price, and since there, for all intents and purposes, there is also no leasing, used vehicles are just displayed with their selling price.

Next is Edmonds.com and KBB.com, both serving a slightly different purpose. KBB.com (Kelley Blue Book) is known as the used car pricing provider, so their main focus is on pre-owned vehicles. Today they offer a twist on the new car pricing by taking the dealers inventory feed number and then using assumptions, calculates what it will cost you to own the vehicle over 5 years. I don't know a single person who has used this, and don't think that the guy who spends $17,500 for a car today wants to have his face rubbed in the fact that over 5 years he will have spent $34,000 for it. Edmonds.com has always been a car information site that also does a good job with used vehicles too. Today, they offer what they refer to as the "Edmonds Price Guarantee" which means on select offers, the price, purchase price, is locked in on that vehicle, but that is not for lease and finance offers.

That brings us to Trucar.com, and they are another animal all together. Trucar.com has a very contentious relationship with most car dealers that it has done business with......they, for the most part, are not dealer friendly. They basically do their best to drive the price of a vehicle into the ground. The primary stockholder in the now publicly traded company is USAA insurance. They drive most of the traffic to the site and have a vested interest in getting the price down as low as possible. They routinely will tell a dealer that he is losing business to the local competitor because he is not pricing his cars low enough. This is fact, as some of my advertising clients do use the site.....most don't, but some do. I personally know dealers who will never use Trucar.com because it is hard enough today to make a profit, so they don't want to work with someone who is driving their already low profit deeper into the ground. Trucar.com gives you a worksheet that you must take to the dealer, with items to add to the price, and take away from the price, plus, you must, just like with the rest of the 3rd party sites give up your information before getting attacked by a multitude of dealers all trying to get your business. Trucar.com is not VIN (vehicle identification number) specific, so they give an estimated price, and estimated savings off of MSRP. In the case of eznewcar.com, we tell you the exact savings on this particular vehicle because we are one hundred percent VIN specific. The shopper sees the MSRP, the selling price and rebates, it is very easy to do the full calculation as see exactly where you stand, not so with these other players. What is also interesting is that these other websites have identified the same needs that eznewcar.com addresses, but have not been able to successfully integrate them into their formats. Either they don't fully understand what changes have occurred in the shopper's attitudes, or can't work the changes into their platform, or more likely the case, they have so much invested in their site and processes that a major change in format to a display and comparison system like we possess, would disrupt and create too many problems. Why they haven't changed isn't the issue, the fact is, the shopper wants these features and with the technology today, it can be offered, if you have properly designed your system

And that leads me to another important point, eznewcar.com, has been designed to be a cloud based system, so we can handle hundreds of thousands of calculations and operations simultaneously. We have used the most advanced software architecture and processes, so we are light-years ahead of what other platforms are running. Our system resides on an Amazon cloud server system, so we have triple redundant security and protection from hackers, man-made disruptions and hosting issues. In a nut-shell, we have taken every precaution to ensure the stability and security of our software.

So, how does it work?

Eznewcar.com is completely free to the users, so they have no reason to not check it out....... The shopper logs into eznewcar.com and comes to this Home Page or the shopper will access our pages through the newspaper web-site, either way the search is "credited" to the newspaper portal so they get complete credit for that customers views.

From the home page, they see the three choices of how they can search for the new vehicle they are interested in. But on the newspaper site, as shown above, we have added a 4th search type and that is for Pre-Owned vehicles. Newspaper need a complete search mechanism that includes Pre-Owned vehicles to sell to dealers, so Pre-Owned is very important. We currently get the pre-owned vehicles but suppress them for the parent site but will allow them for the newspaper portals. Search method one, which should be the most popular, is by price. Also notice on the bottom of the page "featured" vehicles scroll by automatically. Each dealer on the site can choose the vehicle or vehicles he wants as featured. On the home page this is a random scroll of all the dealers featured vehicles, but as the shopper narrows his search to a specific dealer, the featured vehicles are limited to just that dealer's selections. As you see, they have a choice to shop by monthly lease payment, monthly finance payment, and both payments or by the purchase price. This is the easiest and most direct search method in the industry today. By simply using the slide tool the shopper selects the price range he is interested in.... lease and/or finance or by a full purchase price. From there, in all cases they can select the price range they are interested in, for example, in the case of a lease payment; the shopper can select the high-end number, so he can limit the search by his anticipated monthly budget amount.

Once selected, that returns to him all the vehicles that have a lease payment no higher than his selected limit. Now, in the next version release, there will be filters incorporated that will allow the shopper to limit his search by vehicle types, for example, if the shopper is interested in seeing all the, say.... minivans, in his price range, he selects that filter, or the filter for SUV's, sedans, coupes, etc..... The shopper sees all the vehicles in his price range where he can compare them for a variety of items, such as MSRP, cost, selling price, lease, finance price, equipment, any down payment requirements, rebates,

MPG and all the other information shown.

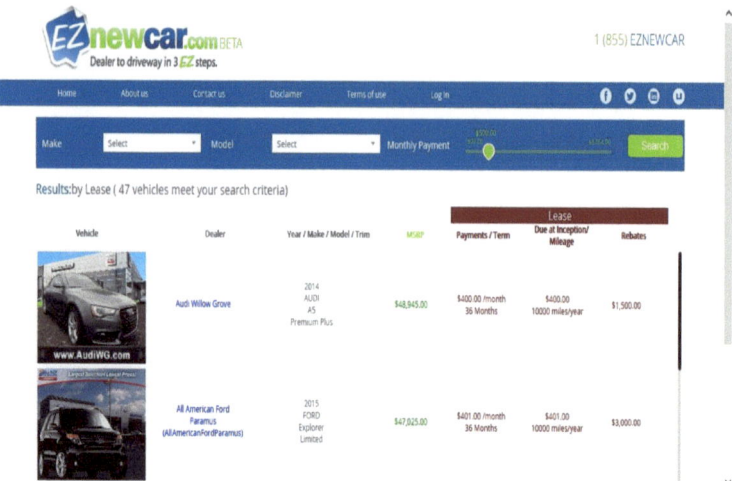

Once the consumer selects the vehicle he or she is interested in, they can select "Contact Dealer" where then, and only then they can send their information to the one dealer they have selected, actually, they can select as many dealers as they like to send info to, but the idea is that once a dealer is selected, the consumer can decide who they want to deal with, rather than having their information sent to multiple dealers who will then hound the shopper to death.

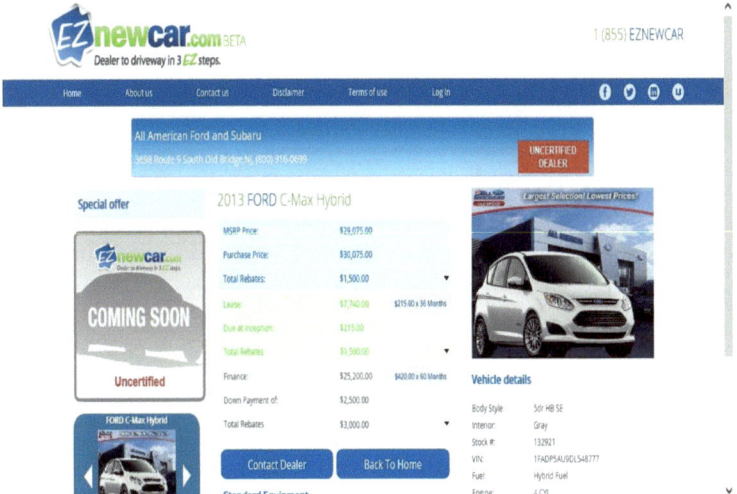

If the shopper decides to select "shop by dealer" he then gets a director of the dealers in his market area.

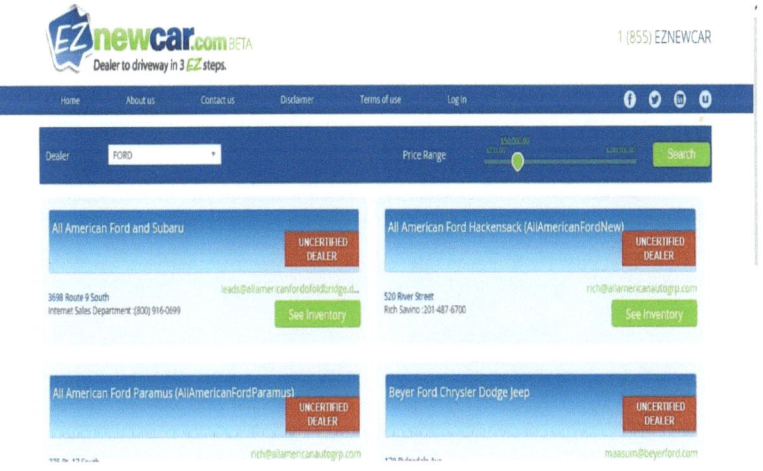

From here he selects the dealer of his choice and clicks on his name.

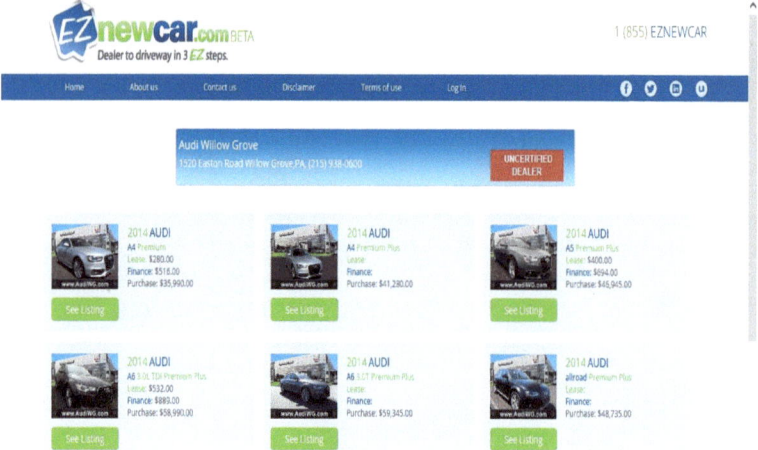

And he sees all the advertised deals that the dealer is offering, and these offers have the best purchase, lease and finance prices which are updated daily.

If the shopper wants to search by a specific vehicle, he just clicks on that icon, and selects the vehicle of Choice

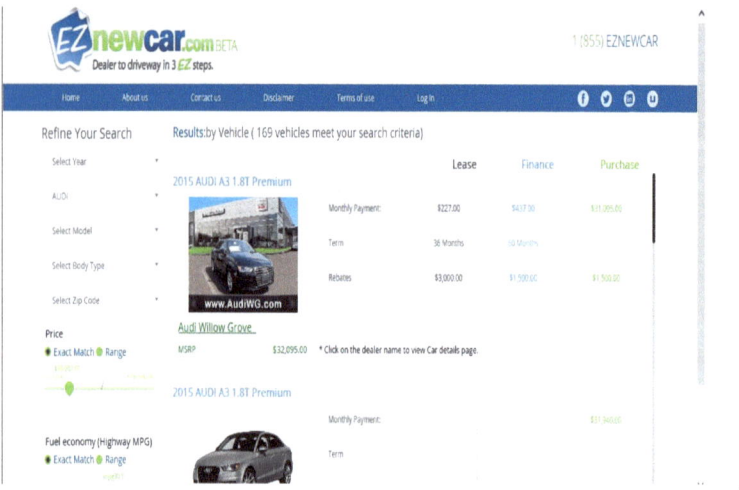

Now he sees the vehicles that correspond to that search listed on the page with the 3 different prices on that vehicle......lease, finance and purchase. Once a vehicle is selected regardless of the way he gets to the vehicle, it always brings the shopper to the detail page for that vehicle and from there, and only there, the shopper can contact the dealer with his name and information with that specific vehicle attached to the correspondence. It is neat, clean and simple, and the shopper does not have to be concerned that he will get multiple dealers contacting him endlessly......just the one or two dealers he initiated the conversation with.

In the next addition there will be another feature built in that will make this system even more superior to the others. Once a vehicle and pricing are selected, in the next version before contacting the dealer, the system, with some additional information from the customer will be able to recalculate the deal exactly to his credit score so that there will be no surprises with the price. Typically, advertised offers are calculated at a 700-score credit rating. If a customer suspects that his credit is not that high, he can have the deal recalculated at his score to see what his payment will be. It will add in sales tax and estimated MV to the monthly figure so that he will know exactly where he stands. As I said, this is the most advanced search and pricing web portal platform in existence today....and therefore I say we will change the way people buy cars, most likely forever.

The second method of search is by vehicle, which I think will be the second most popular method. A shopper who selects this method need only select the brand of vehicle, Honda, and on the second drop down select the model of Honda they want to see, say Accord. Once clicked, the shopper sees the Honda Accord vehicles arranged by price order. He can scroll down, find a vehicle he is interested in, and view the complete pricing amounts, lease, finance and purchase. Once again from there he can contact the dealer, who get the email, with that exact vehicle tagged to it and begin the process from there. A copy of that information is held by eznewcar.com for future analysis, but will never be given out. We will compile the stats, but never sell any of the shopper's personal information. And like the first search method, once the next generation software is available, the customer will be able to fine tune the monthly pricing to his credit rating.

The third and most likely least used method of search is to search by dealer. Here the shopper will see a directory of the dealers selling that make of vehicle in his region……so if the shopper selected Chevrolet, he will see a directory of the Chevrolet dealers in his physical search area. From that directory, he can select the dealer and then see the vehicles, segmented by models that are being offered for sale. This method most closely resembles the newspaper format of the Sunday ad for the dealer with all his best offers on one page. Once he selects the vehicle he is taken to the final sale page for that vehicle. In all searches, the search terminates at the same page, which is the vehicle ad page that has all the specific information on that vehicle and has the means to then take the shopper to the button to "contact the dealer". The focus of eznewcar.com is to take the vehicle information provided by the dealers pricing selections, present his best offers and then have the shopper contact the dealer to arrange the sale. Eznewcar.com does not sell the vehicle, but is the conduit that expedites the process so that the consumer has a hassle-free shopping experience and the dealer ultimately gets a very highly qualified sales lead. As I point out in another section, internet leads usually are a relatively poor lead do to duplication and timing in the buying cycle, but because eznewcar.com leads come to the dealer when the shopper is looking at the pricing and is ready to make that purchase decision, our leads will be much higher quality, and worth more to the dealer.

As eznewcar.com grows into additional regions, shoppers will have the opportunity to view deals and dealers from those additional regions. He need only select the region from a drop-down menu, select the one he wants and then view the same way as he would in his own region. The value of this feature is that if someone is moving to another region, say from New Jersey to Florida or Texas, they can compare deals in both regions and decide of where to buy if they are planning a new vehicle purchase either before or after the move. Just another example of how this website is tailored to the needs of the consumer.

How did this come to be?

So, the basic concept grew out of my desire to take the Sunday newspaper auto classified section ad and incorporate it into the internet where we could use the superior performance characteristics of the computer to give the consumer the ability to compare and modify the deals offered by the dealer. What sets eznewcar.com apart from the other 3rd party vehicle display sites are twofold, one, we designed it to replicate the traditional newspaper ad showing the best price as advertised, where the other sites replicate the dealers web site, and two, the shopper gets the information they need anonymously so they can shop in secure confidence knowing that they will not be bothered unnecessarily.

Pricing, I determined should be on a monthly subscription basis as the newspaper model it was patterned after, the only difference is that eznewcar.com has a monthly subscription and newspapers have a per ad subscription. Further, I watched as Trucar.com struggled with a performance based pricing format. Trucar.com charges a $299.00 per new car sold, and $399.00 per used car. The used car portion was not relevant, but I watched in the public (auto industry news reports) heated back and forth between Trucar.com and the dealers fighting tooth and nail for" who's" lead it really was. After all, one of the mainstays in the retail auto community is the referral customer, or better yet, the repeat customer. It is well known that they cost the dealer far less to cultivate and sell to because they have previously purchased or were referred by a previous customer. The dealer could argue that he sold this guy 3 or 4 cars in the past, and now on this purchase the guy looked at the Trucar.com site, registered with them and so now when he went back to the same dealer he brought from before, why should the dealer pay Trucar.com $299.00 for the privilege of selling him again? This format has created a very contentious relationship between them where Trucar.com was forced to write off millions of bad debts in the past, and now attack the dealer within minutes of a deal expecting to be paid on the spot. Neither side is very happy, so that is why I resisted going down that path. Next, I wanted to make this, at least in the initial market, attractive to all dealers regardless of their size and sales volume. We have come up with a method where we match the monthly cost to the dealer's sales volume, the cost per month is based on the average monthly sales volume for the 6 months prior. In this way, the cost is far more agreeable to the smaller dealer who would have a hardship absorbing a higher priced model, and as I said, this is for the initial NJ market where we can test the effectiveness. Personally, I believe we are very underpriced for the service we will provide, but we need to work out the issues early so when we expand we have a smooth product transition. Therefore, the subscription method we chose made perfect sense to me since the Newspaper advertising was slipping away, I knew that the consumer would want to find a way to get the information that the newspaper once provided, but in a different, expanded and more complete way. The original concept called for the dealer to provide the information monthly, as he did to the ad agency so that a new ad could be created, and his best deals would be presented. Now, I knew as well as anyone who deals with car dealers that they for the most part is lazy and difficult to get information from, even when it will be a direct benefit to them; they would rather make demands of others and expect others to do the heavy lifting for them. That is a given, but even I was sadly mistaken. I expected that I would get a minimum of co-operation from the dealers, but that was not meant to be.

Because of this issue, we were forced to take a step back and re-evaluate the site and the process. It became clear within a month that dealers would not help get us the information we needed to make eznewcar.com as effective as I knew it could be. Herein lays the irony of the current internet 3 party web sites and why eznewcar.com will be far superior to all of them. The current 3rd party sites take a passive approach to gathering their information. Cars.com and Autotrader.com, for the new vehicle side of their site, remember, eznewcar.com is solely directed to new vehicle business, take the dealers website feed and present that as is to the public, the only difference is that they intermingle the same kinds of cars from competing dealers, but for the most part at MSRP, and should the dealer supply a different pricing model, the consumer will have a search result that is basically the preverbal " apples and oranges" where different numbers are being displayed further confusing the consumer. Edmunds.com and KBB.com basically are guilty of the same approach, but they put their own little spin on the display, like a "guaranteed price promise" or "the 5 year cost to own", but again, this is based on MSRP or a number similar and not actually what the consumer can expect to pay, especially if they are a payment buyer, and today the statistics show that around 90% of vehicle buyers are lease or finance buyers so to them the most important number is "what is my monthly cost, and does it fit my budget?". So, it became obvious to us that we better change it now before we get any deeper into the project so that we have the best product we can. Another issue we encountered was that we had no way to remove sold vehicles quickly and efficiently, so we know changes needed to be made. It forced us to basically shut down our sales and put everything on hold until we could reconstruct it the way we wanted it. At this point I want to point out that we saw an error in our thinking and made the decision to freeze everything for a few months until we could get it right. As a company, we are committed to put the best product possible out to our customers regardless of the issues, because it is our reputation and future that is at stake and we will not leverage that for anything. This, I believe speaks volumes as to who and what we are…. we do not take this venture lightly, as we have put countless man hours and our own money and sweat into it. We are fully committed.

In the first version we took the feed from Chrome Data, a highly respected company used by most in the automotive world to take a vehicle identification number (VIN) and decode it into the vehicles information so we could display it. As this process worked well, it did have many issues in the vehicle style codes and in the cost calculations as it missed a number of the options in the vehicle so that created a lot of manual labor on our part, which it was clear, was going to be a problem going forward as we added higher volume. And even a larger problem surfaced as it became increasingly difficult to get the vehicle pricing and rebate information from the signed-on dealers. The dealers had been spoiled by the other 3rd party websites that had taken their information from the dealer's inventory as displayed and that required no work on their part, even though the dealers complain about the lack of effective leads, they have shown no interest in helping their own cause. We knew we had a problem and needed a rapid and complete cure. We did our research and found that with the integration of the services of HomeNet, based in Pennsylvania, we could pull the complete new car inventory of dealers complete with pictures, MRSP, complete detail of options and standard equipment along with cost and selling price information. We now have access to over 2,100 dealers' inventory nationwide and 140 dealers in NJ alone. We opened up the revamped system with over 18,000 vehicles just for the NY, NJ metro area. This solved the first problem in it provides the dealers inventory where we can display each type of model vehicle with the lowest priced vehicle from that range. Dealers, who are not previously signed on with Home Net, can be signed up by us as a "light" dealer for $40.00 dollars and we will have the necessary info we need for them. Another huge advantage of using HomeNet, we can easily set the pricing parameters for each model vehicle sold by either a dealer selected dollar mark-up or mark-down, depending on his desire or by using a percentage mark-up or down. This method allows the dealer to have complete control over his sales display and it allows him to change it as he desires.

This solved our first main issue, now we needed to correct the second flaw in our programming. This issue is caused by the need to display the proper lease and finance numbers to really set us apart. In the initial format, even though it was a temporary fix, we used an ADP Desking software tool to calculate these numbers. The system works, but because it is a "generic" system not mated to any one franchise, as it would be in a dealership, the system was difficult and clumsy to operate, and because of that, it was very time consuming. All of which was unacceptable, but this was never our end game, only a stop gap solution until the correct system could be installed. All along, that system was planned to be the "Drive it Now" software from a company in Ohio. After a number of meetings, countless phone calls and even a contract, it became clear that they had "over promised and under delivered" on what they said they could do for us. We were forced to look elsewhere to get what we need. We were fortunate to hook up with DealerTrack, one of the largest providers of a number of integrated dealership solutions. One of these solutions will plug into our software and provide the necessary lease and finance calculation so we can display each vehicle with the purchase, lease and finance pricing along with the necessary vehicle information and picture complete with the regional rebates, so the final pricing can be very accurate.

Further, to the software internal changes we made, we gave the website a new and softer look. In marketing we all know that babies and pretty ladies sell very well. We adopted a softer look with a less aggressive feel to it. When many "looks" were shown to a group of focus group participants the new look was favored 3 out of 4 times. Because of the 75% approval rate, we opted for the new look.

To recap our process now, each night we bring in the updated dealer inventory in a batch so that we have eliminated the problem of removing sold units from the display. In the dealership as the unit is processed as a sale, it is removed from their inventory database and therefore, it does not come to us the next day. We track the vehicles that fall out of the system to be cross-referenced with our shopper's views to help us gather important data on the market activities.

A number of people have suggested that we sell banner and skyscraper ads on eznewcar.com. Although that is a potential source of additional income, I have resisted that idea for a number of reasons. First, the consumer does not need to have to navigate around pop up and banner ads.... I know they bother me when I go on other websites. For example, go to weather.com and just try and find the weather. There are ads popping all around you, videos about all kinds if unrelated nonsense, fish at the bottom of the ocean, sharks swimming up rivers, just the most ridicules stuff you can imagine. Go on cars.com, look up a brand of car, and on the same page of the car and dealer you are looking at, there are ads from the factory for competitor's brands of cars. I know, if I owned a dealership, paid to be on the site, and because some other dealer with the same, or different brand paid more, he gets an ad on top of my page......that would have me up and angry. Next, those ads slow down computers because of the video and action they have; I do not want anything slowing down our shoppers when looking for a deal on eznewcar.com. I want a level playing field here. I do not want one dealer getting a better position or more exposure then the rest; all dealers get the same treatment, as that is the proper way to operate the site. On eznewcar.com the deals do the talking.

Earlier I mentioned that eznewcar.com grew out of and has its heritage rooted in the marketing processes of the very beginning of auto retailing. In the beginning there was newspaper advertising for auto dealers.

Here is an actual newspaper ad for DeCozen Chrysler in Newark, NJ that dates back to 1926. The newspaper advertisement has always been the cornerstone of all auto dealers marketing because it spoke to the essence of what the consumer was most interesting in, the cost and features of the vehicle. Granted, there are many other factors that influence the decision to purchase a new automobile, brand reputation, vehicle size, features, MPG, safety ratings, functionality, dealer reputation, even color, but at the foundation of any purchase decision is the price.....the cost to the end user. In the beginning, the price was easy....the cost was represented only as a purchase price; fore we had not yet invented the finance and lease variations to confuse the buyer's ability to make a simple decision yet. But today, there are a multitude of factors and finance options, coupled with rebates, subvented finance rates and additional down payment and lease annual mile options that the process has become challenging to even the most astute shopper.

There are a number of surveys done that show that newspaper advertising *was* the most effective means to reach auto buyers in the most critical portion of the buying cycle......the final two weeks. These surveys showed that cable advertising was most relevant in the 2 to 3 months prior to the purchase date, but fell off dramatically as they were effective in branding a dealer, but did not do an effective job of making a strong "call to action" in their message so as to sway a shopper's decision. Radio and billboard have been shown to maintain a flat effectiveness throughout the lead up to the purchase, 3rd party websites have also shown that they peak 2 to 3 months out because the shopper is , at that point, just gathering data on

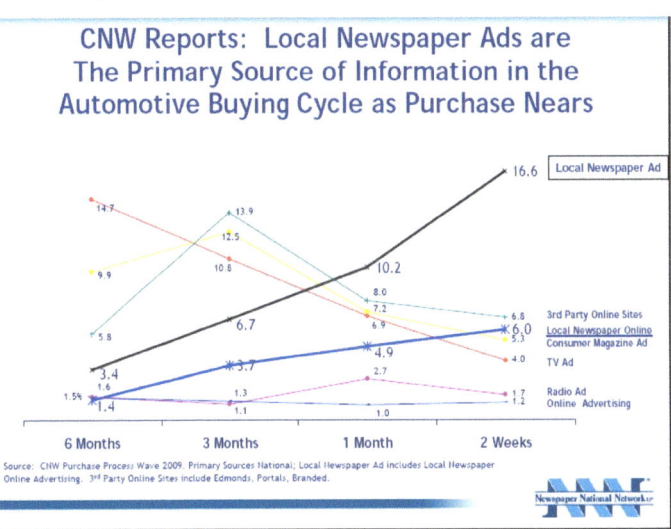

the vehicle (electronically, "kicking to tires, so to speak"), availability and general MSRP relevant to the other vehicles in that marketplace.

This CNW report is clear proof that shoppers are searching for pricing information just as they are ready to purchase, and that is what eznewcar.com is all about

Manufactures and dealers websites attract attention closer to the buying point, as now the shopper is looking more for availability and location, more than any other factor, but because the information is mostly "structural" and not price oriented, the shopper was still forced to look to the newspaper ad to find the best lease, finance and purchase price information on vehicles he was interested in, but because it was "scattered" at best, it was a time consuming process looking through a number of newspapers and still not seeing all of the dealers represented or even, quite possibly the vehicle the shopper is interested in. And to complicate this situation even more, newspaper viewership has fallen dramatically due to most people getting their information real time from 24-hour news networks, radio and now the internet. The advent of the real time internet reporting and interaction has driven the news information consumer from day old newspapers, to the current, up to the minute news cycle of the internet. And along with the migration of the news consumer away from newspapers, this has prompted the migration of the auto shopper to the internet too. If the shopper is looking to the internet, well the dealer has to change his thought process too. So today, the newspapers have been forced to lower the price of their advertising platform to dealers (more on the effect of that later) but even bargain basement pricing hasn't slowed the defection of the dealer away from newsprint. Since newspaper marketing was the mainstay of auto dealer marketing, the agencies who handled that medium were also devastated by the loss of income too, as mentioned earlier.

Surveys have shown that anywhere between ***85% and 92%*** of auto shoppers touch the internet during the shopping process, depending on whose survey you view, but personally, today, I think it is impossible to look to purchase a new vehicle and not, look at the internet at some point. On top of that, consumers send between ***7 and 11*** hours, again depending on the source you contact for that information, researching the purchase of a new vehicle. But still other recent surveys by none other than USA Today, along with many 3rd party internet sites (Cars.com, Autotrader.com, Edmunds.com and Trucar.com) uncovered the nasty little secret of internet auto shopping. The survey and report showed that still with all of the sites with information available to shoppers, the shoppers were more confused than ever. The information tends to contradict itself depending upon the source, the pricing models were varied and could easily show different prices on the same vehicle, further, and even more disturbing to shoppers is that they have to identify themselves to internet shopping sites before they get any pricing back from the dealers, and ***73%*** of shoppers surveyed said that they will abandon the site if they have to offer their contact information before getting what they want.

> *"73% of online shoppers say they will abandon a website if it requires the shoppers information before allowing the shopper to find what he wants……..Google Zero Moment of Truth Survey*

Think about that, I am willing to say that you are mostly of the same mindset. I know I will not give my information blindly......I do not want to be flooded with phone calls and e mails from overzealous sales people who will stop at nothing to either sell me a car, or be able to convince his manager with a log of all of the attempted contacts he has made. No thank you and this is one of the reasons why car shoppers are so dissatisfied with the process today. When surveyed, consumers are quick to point out that they hate the system as it is today.

SAY THEY ARE "MORE LIKELY" TO BUY A SPECIFIC MODEL OR BRAND IF THEY FIND POSITIVE COMMENTS ON SOCIAL MEDIA

ARE WILLING TO BUY A VEHICLE ONLINE

About the Report

Cars Online 2014 is Capgemini's 15th annual survey.

All respondents were "in-market": 94% were planning to buy a car in the next 12 months. 79% would be purchasing or leasing a new car, and 16% were in the market for a used car. 5% had still to decide whether to buy a new or used car.

Capgemini worked with ORC International, a global research firm, to conduct the survey for Cars Online 2014. All analysis and interpretation of the data was made by Capgemini. Fieldwork was conducted in February and March 2014.

In a world-wide survey done in 2014-2016 (USA shoppers were a major contributor) **44%** said that they are ready to buy a car on-line right now if they could. In a 2014 Edmonds.com survey, **54%** of the respondents said that their "biggest Un-Met need" was not having an out of the door price before going to the dealership. Look, I can throw around numbers all day, but the bottomline here is that the current situation is highly flawed, and consumers and dealers want a better way, to buy and sell cars.

Take a minute and review what I just said, and think of how these issues and problems can all be corrected in one comprehensive well-thought-out process. Dealers have lost their most effective means of "price" marketing, the newspaper. Marketing of dealer's inventory has migrated to the internet, but the internet has muddied the waters with inconsistent dealer information, pricing and vehicle information. These sites require consumers to offer up their personal information before getting anything back from the dealer, and then it is an all-out attack of e mails and phone calls doing everything they can to get your business while pushing away your business. Consumers routinely say that they hate the process because the dealer confuses them with a maze of information and well devised sales tactics designed to force the shopper to "cry uncle", just cave in and buy the car just to put the process behind them. The fact that nearly half of the shoppers today would buy a car completely on line if they could, speaks volumes about how much they hate the negotiation and buying process as it is today. The fact that dealers consistently push the shopper into the test drive is because they know that once you get into the brand-new vehicle, with all of the new technology, tight fit with no rattles and new car smell, that the shopper is sold, but the truth is.... today, all of the cars are the same. All minivans drive the same.....if you could drive them blindfolded, you could not tell them apart....Honda Accord, Nissan Altima, Toyota Camry, Ford Fusion, Chevy Malibu.......they are all the same, in fact, most of the components in the vehicles come from the same sub-contractors, so there is really not enough difference to matter, but dealers want you to buy their car, so they push the test drive to get the deal closed. And most importantly,

> *54%* of market shoppers claim their biggest UnMet need is not having an "actual out of the door price" before going to the dealership   -   Edmonds.com 2013

there is no place (this is where the newspaper excelled) where the shopper can go and find a vehicle that fits into his or her monthly budget quickly and easily while still maintaining anonymity. Every one of these issues and so many more are resolved to everyone's satisfaction with eznewcar.com. Clearly a case of "one size fits all".

Last week while interviewing a potential auto dealer marketing client, they revealed that their typical monthly advertising budget was roughly $70,000. That is a sizable amount of expenditure, by any dealer's standards for a single point single brand dealership.... of that, about $8,000, or just 11% of the ad budget was allocated to print, which is down from the heady days when about 60% of the budget was tagged to print, about $15,000 spent on cable and the remaining $47,000 went to some direct mail (losing effectiveness rapidly) but the lions share going to a variety of on-line marketing sites. Autotrader.com alone accounted for $6,500 per month which, in my opinion is way overpriced for its effectiveness. Around $20,000 a month went towards targeted and re-targeted email blasts to reach people who may, or more likely, may not be in the market for a new vehicle. The point of including this information here is to point out, that there are no shortage dealerships spending sizeable amounts of cash on marketing and that the largest share of it goes to on-line sources. This dealership is typical in its spending habits......although many spend less, and some spend more, the percentages of on-line marketing to conventional marketing remain fairly consistent, on-line is where the action is, and dealers will spend whatever they have to get the results they need.

In this pie chart below from NADA compiled information from actual dealers, from 2012, which is the most recent information available, the newspaper advertising has fallen from 47% in 2002 all the way down to 17% in 2012, and it has fallen even more in the past year. But Internet marketing has grown from 5% up to 27%, and that has grown even more in the past year. It is clear where the trends are going and eznewcar.com is the only web based portal that will deliver accurate selling price information real time to

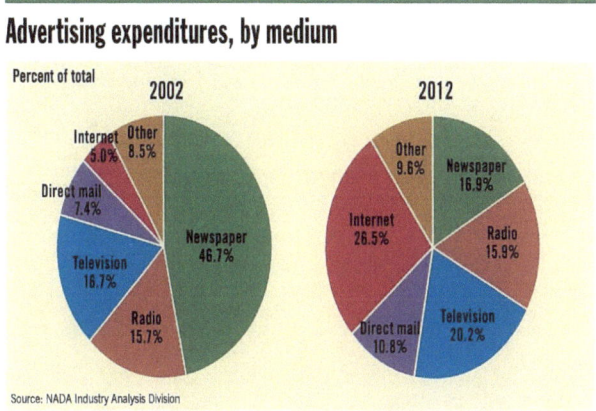

the shopper.

Another strong support to the eznewcar.com concept is that Ford Motor Company which pays advertising co-op to its dealers for approved advertising made a major change to their pay-out program this year. In the past, any approved advertising (must be pre-approved to qualify) was eligible to receive .50 cents on a dollar spent back in co-op reimbursement and most internet marketing did not qualify. In January of this year, Ford announced that effective July 2014, a full 50% of all marketing done by the dealership must be digital, that is quite a shift in policy, going from nearly no approved digital marketing to half of the advertising budget. Clearly Ford understands the importance of digital marketing and put a tremendous value on it. So Ford and Lincoln dealers must shift to more digital marketing and Chrysler is following suit, so GM will not be far behind.

And if that is not enough reason to highlight the importance of the need for a website like eznewcar.com, AutoNation, the largest retail consolidator of dealerships announced to all of the 3rd party website providers this past April that AutoNation would be pulling out its 229 stores from their programs, unless they made major changes because AutoNation believes that they, the 3rd party providers (Cars.com, Autotrader.com, Trucar.com, et…) charge far too much for the rather poor leads they provide. Now, I haven't touched on that point previously, but will here. One of the inherent problems with the current batch of 3rd party lead providers is that they tend to push leads to the dealer that are not ready to purchase yet……if you will recall, earlier I pointed out that 3rd party lead provides shoppers who are nothing more than "electronically kicking tires" these are shoppers who are gather info on vehicles in preparation to the future purchase of an auto, but the dealership pays for that lead, so they will contact him and push that as far as they can in an effort to convert it into a sale….this leads to a poor closing ratio for on-line leads and to the consumers frustration with the current system as it is. Further, since shopper today routinely shop multiple on-line sites, the dealer frequently gets multiple contacts for the same lead, which also hurts the closing ratio and further upsets the

> **Jackson** (Michael Jackson CEO of AutoNation) **to 3rd party lead providers :** "We'll get our own leads! As seen in the Automotive News, April 21, 2014

shopper. Remember, the industry surveys clearly show that 3rd party websites, since they are more "information based, and less price based" have their peck activity centered around the 2 to 3 month prior to purchase time, so these are not "hot leads" of people ready to take action. Back to the closing ratios I mentioned. In a typical dealership, a" walk in", or "call in" customer who saw an ad, or has decided to purchase a vehicle is "closed or sold" roughly 4 out of 10 times, a *40%* closing ratio, but because of the poor quality of the internet lead, because of the duplication of the lead, or the timing, well in advance of his true purchase period, these leads tend to get sold or closed 1 out of 10 times, for a *10%* closing ratio. Therefore these leads tend to cost the dealership more, but have a much lower closing ratio. That is the main reason why AutoNation sent the letter that they did in April of this year to their 3rd party lead providers. AutoNation was very direct in their letter, they told these providers, either fix their broken systems, which they can't without a major scrapping of their system and adoption of a new system similar to eznewcar.com, or substantially lower their rates to be more in line with the quality of the leads they provide or AutoNation will develop their own system. Now that may or may not happen, but the message is clear, if the largest retailer of new vehicles in the USA is unhappy with the current crop of 3rd party providers, then many others are unhappy too, but since they don't have the deep pockets of an AutoNation and/or the clout of 229 dealerships behind them, what can they do? Well for starters, eznewcar.com will provide them with a state-of-the-art lead provider shoppers /buyer system for far less than the current guys charge.

As little as 5 years ago, a process this simple and effective could not be entertained because technology and shoppers and dealers mindsets were in a different place then they are today. But because of the shift in attitudes, and the explosion of smaller internet viewing devices, such as smart phones, tablets and mini-laptops, internet shopping has reached astronomical levels, and there is no stopping it now. Shoppers today have in their hands at all times the means to verify all the information they are seeing in the store, and that goes from the price of frozen corn in the supermarket, or bedding in the discount retail store to the price of gas at the pump. In a survey completed just this past summer, by the Automotive News, it was shown that **63%** of auto shoppers used their smart phone or tablet while on the lot of a dealer to cross check information that they were seeing real time and of those, **56%** looked at 3rd party sites, not the dealers site, so clearly, 3rd party sites get great selling traction. This survey also highlighted the fact that **trust** is a major concern of the shoppers, now as that should come as no surprise to anyone who knows the car business, but where the shopper places his trust may. The number one most trusted website category is the 3rd party site, again, consumers tend to place more trust on the 3rd party site then even the manufacturers' or search engine dealership ads or even the dealers site.

In a recent July 2014 survey done by LeadiD.com, an excerpt shown below, dealers and manufacturers web sites once again took a back seat to the 3rd party websites, which is the case in all surveys. Yet, there is surprising results on issues that this survey makes clear. First, Kelley Blue Book leads the pack primarily because of their expertise in the pricing of used vehicles and because they advertise themselves as the "Trusted Source" so obviously, they have done an excellent job of marketing. The vehicle information on their site is no different than that of other sites and pricing is no different as they all take the dealers website feed to establish the price, the main difference here is the "5 year cost to own" piece which means nothing to most people and absolutely nothing to "lease" buyers. And with only 32% of consumers and about 16% of dealer's approval, these are not overwhelming endorsements of complete trust. Interestingly, dealers like the idea of autotrader.com more than other sites, but consumers place it in the middle of the pack, along with cars.com and Edmunds.com so there is a disconnect there, but what I find most interesting here is the complete omission of TrueCar.com from both sides of the survey. I am no fan of TrueCar.com with their "bullying" style of approach but even I was surprised, and pleasantly surprised to see that they did not even place here. This validates my thinking better than I could ever evaluate it to you.

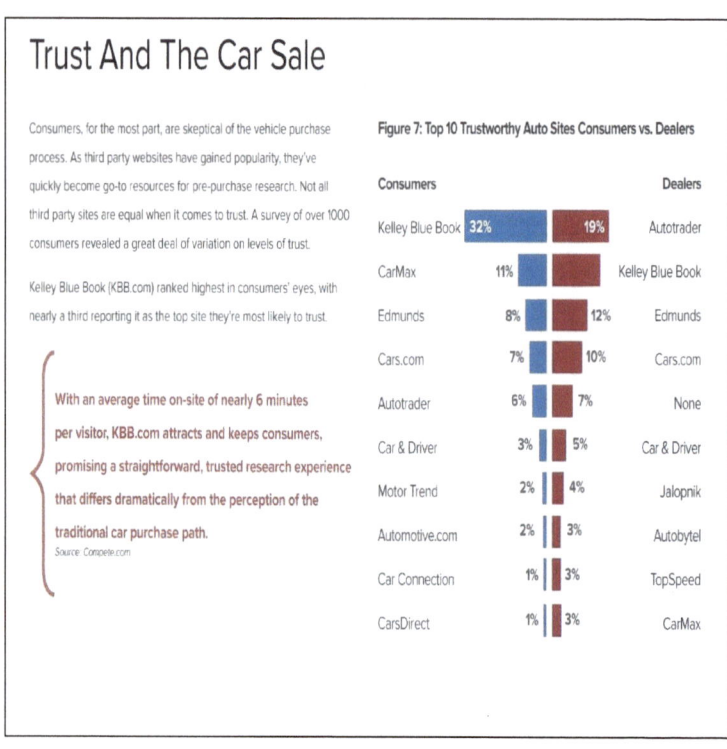

And secondly, here is a chart that tells another interesting story, also from LeadiD.com.

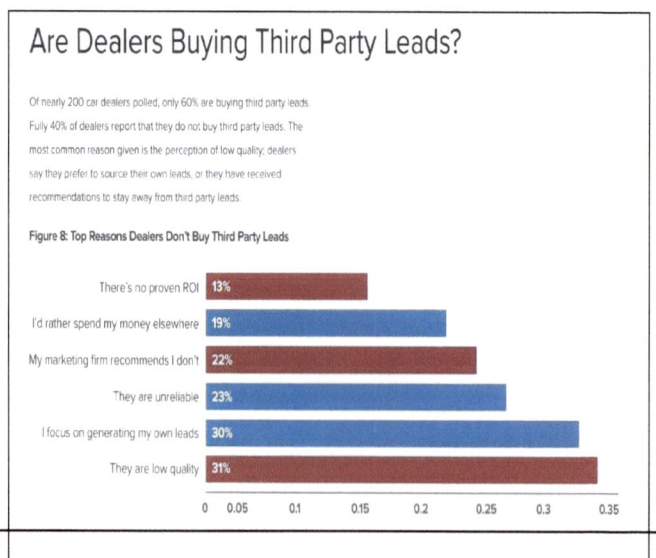

This graph points to a few very important points which will help eznewcar.com stand head and shoulders above the other 3rd party sites. Remember, this information was gathered in July 2014 and just published this week. Let's start at the top and work our way down. First, with eznewcar.com the return on investment (ROI) is very easy to calculate each month….dealer pays a price each month based upon his average sales per month so the small dealer is not hurt by a one price across the board policy and because he gets a complete dashboard he sees all of his leads, where they came from, what vehicle they were looking at and what pricing model drove the traffic, all necessary information for the dealer to use in future sales, ordering and promotion decisions, Secondly, sure go ahead and spend on other marketing platforms, I would encourage that as no one source is going to get all of the leads, but again, our volume based pricing means that a dealer can incorporate eznewcar.com without disturbing his other marketing too much. Third, as mentioned elsewhere, we can cut the advertising agency into the commission cycle with eznewcar.com unlike the other 3rd party sites, so they will not work against us, but actually promote us. Fourth, we will be very reliable for three solid reasons, A., we update every night so sold inventory is removed, only fresh available units are displayed, B. we update the deal every night so rebates, finance rates and lease residuals are always the most current up to date available and C, we do not share our leads with other sites as the others do, so when a customer comes into eznewcar.com and reaches out to a dealer, that is the only lead that is generated from this interaction, so there is no duplication or multiple lead contacts…..just this one and this should be a serious buyer. Fifth, fine, let the dealer generate his own leads, but again, at our low cost to the dealer, he can afford to use us in conjunction to whatever else he chooses, and lastly Sixth, and the biggest reason against 3rd party site leads is and has always been their low quality. Here is where we shine very brightly…….our lead is a solid lead ready to buy in just about all cases.   Let's think back to the chart I showed you earlier, the chart that showed why shoppers always looked at the newspaper ad right at the last minute before deciding on what and where to buy his vehicle. And that fact is solely because the newspaper reviled the price, purchase, lease or finance; it has always been the newspaper that presented the price. Websites for the longest time only gave up the MSRP price but nowhere could the buyer compare competing dealers' price on the same model other than by MSRP until eznewcar.com.

Shoppers have a very cynical eye for the car dealer, most well earned, but some of that mistrust is just urban legend, but the fact that consumers trust the 3rd party site the most means that eznewcar.com will instantly have a leg up on getting the shopper to look at it. Now a good portion of the on-lot research was on used vehicles where pricing was being verified, but none the less, shoppers are showing that there is no limit to how far they will go to be empowered in the auto sales process. For all the years leading to now, the dealer was driving the entire process, and now consumers are finally realizing that they have access to the power also and they will not relinquish it in any way…..in fact, they want more. And the most important piece of information is a monthly lease, finance price or the actual purchase price after applicable rebates. But up until now, the ability to see the final "out of the door" price on a new vehicle has been elusive, if not all but impossible to see without the long drawn out negotiation process that we have come to hate. Remember, 54% surveyed said their biggest unmet need is not having an out of the door price up front……that is one of the points that eznewcar.com excels at. Eznewcar.com makes that final piece of the puzzle fall into place. I would like to steal portion of the line that Hertz made famous back in the 60's……..*"eznewcar.com puts you in the driver's seat".* The truth is, we do that in a few ways, we empower the shopper with the right information he needs to make the right decision for his personal situation, and actually put him in the driver's seat of the new car or truck he wants.

Business Model Canvas

Business Hypotheses

1. **Data Suppliers:**

    ChromeData: Vehicle Identification Number (VIN) Explosion, provides eznewcar.com platform the capability to list all vehicle information about a given VIN. This information can be gathered on run time and can be validated from the vehicle manufacturer's inventory.

    HomeNet : Inventory Feed that is swept by a dealer management system (DMS). Eznewcar.com platform capability is enhanced to display live inventory present in the dealer's systems using a batch process for processing data feeds on a daily and hourly basis as needed. This capability enables Eznewcar.com platform to maintain live inventory status with respect to every dealer that is signed up Eznewcar.com.

    DealerTrack : First Pencil (initial Price Quote) provides Eznewcar.com platform the capability to calculate an initial quotation as per consumer's data (e-mail address, address, phone number, and interest in a vehicle choice). While a consumer provides minimum information for realizing an initial quotation, this quotation will give a consumer a close to real time deal information for an actual purchase of the vehicle (lease, finance or outright purchase quote).

    White-label solution powered by eznewcar.com: This is a service offering provided to media companies that want to have a "powered by" search engine using Eznewcar.com's REST based services model to power their front end systems for displaying dealer inventory and advertising data. Application Processing Interface or API for short is a method used to provide services that would enable 3rd party websites and applications to retrieve data from Eznewcar.com and display as per agreement for advertising and marketing purposes. This service is applicable to media companies such as:
    - Greater Media
    - Shaw Media
    - 14 additional media companies

2. **Key Service Features:**

    Provide consumers with accurate vehicle pricing and selection

    See Purchase, Lease and Finance Payments
    Shop by Dealers, Make and Model and / or Price
    Shop completely anonymously
    Provide Media Companies with a state-of-the-art platform for auto listing
    Simplify the Shopping Process for Consumers
    Provide Higher Quality leads for Dealers

3. **Unique Product Benefits for Dealers:**
    Roughly ½ the cost of the competition.
    Shared Revenue Model with Media Companies
    The only 3rd party site that allows Advertising Agencies to earn a commission
    No Competing Dealer Advertising on another Dealers Page
    Completely Automated Process, Dealers have Minimum contact

4. **Value Resources':**
    Dealers Inventory Swept Daily by HomeNet
    Lease and Finance Calculations done Daily by DealerTrack
    Regional Rebates are Calculated into each Deal
    Residuals and Interest Rates Updated Daily

5. **Customer Relationships:**
    Personal Contacts from Years in the Businesses
    Ads placed in Daily Clips
    Press Releases Primarily in Daily Clips
    Testimonials from Media Companies On Board
    Sharing of Revenues
    Sales Package, Training and Hands-On Sales Assistance in Start-Up Mode
    Shared Marketing
    Continuing and ongoing Customer Support

6. **Marketing Channels for Eznewcar.com platform:** In addition to signing up with Media companies, and advertising agencies in order to maximize Eznewcar.com service offering coverage, following marketing channel mechanisms were signed up with:
    We Sell Through Media Companies which other 3 rd party sites do not
    "Aggregator" sites typically available to media Companies have no real auto pricing included
    Other 3rd Part Sites Avoid Media Companies
    Consumers can access the account by either the Media site or the eznewcar.com direct
    By using Media Companies Organizations, we get complete auto dealer coverage without an expensive sales force on our books.
    Using Media Organizations spreads eznewcar faster and to more areas
    Rapid growth and extended reach without the need to set up regional sales management

7. **Customer Segments:** So what makes the Consumer and Dealer to visit Eznewcar.com platform?
   We provide an easy to implement plug-in auto search widget for media companies
   No set-up cost to the media
   Any and all dealerships can use the service to drive traffic
   Auto shoppers can eliminate the unpleasant haggle and negotiations from the purchase
   Dealer CSI / SSI will increase due to simplicity of the buying process
   Completely inclusive, no limitations on Media Companies or Auto Dealers in system

Validating Business Model

Overcoming dealer objections to eznewcar.com (and Media Auto Site)

Q. I already do a number of internet marketing website programs, why do I need another one?

A. eznewcar.com is different; it is the first and only site to engage car buyers with pricing information, on a specific vehicle, right down to a monthly payment, without that shopper revealing their info. **73%** of auto internet shoppers will abandon an auto shopping site if they have to give up their information before getting information they are seeking. Eznewcar.com is the only 3rd party site that allows shoppers to search, all the way down to the monthly payment, while maintaining complete anonymity, no one else can claim that fact. And since between **83% and 92%**, depending who you believe, of all auto shoppers begin their search on-line for a new vehicle, you can never have enough exposure on the internet.

Q. What makes eznewcar.com different from the other 3rd party automotive sites out there?

A. A number of key elements differentiate eznewcar.com from the others. First it was designed by former retail automotive and advertising professionals to finally give the car buying public exactly what they want, a one stop shopping experience where many vehicles and pricing options are displayed side by side with the necessary information to make a truly informed decision. *(Built in the image of the Sunday Automotive Classified Advertising Section that dominated the auto marketing world prior to the internet age)* There are no "pop-up" or "banner" ads to distract the shopper, no dealer can purchase a better position and vehicles are set by random selection so no one gets intentionally to the back page......it is a level playing field where your vehicles and pricing rule the day. Vehicles are shown with their purchase, lease and finance price in one place for the first time ever.

Q. *My advertising budget is complete; I can't find room for another product.*

A. There are many reasons why you can't afford to not advertise on eznewcar.com. First, it is very cost effective, at no more than $1,500 per month, and in (your case, only $XXX per month), it is as much as 4 times less expensive than its competitors' monthly cost. Take your least productive marketing source, cut it back by the cost of eznewcar.com and after a month or two, you will see a dramatic improvement in sales. And, remember, the manufacturers co op programs (if they apply to this franchise) can cut the cost of your subscription in half, so this is really a bargain in the world of marketing.

Q. *We don't advertise price on the internet.*

A. Okay, I understand that, but times are changing, **54%** of car shoppers stated that their biggest unmet need before traveling to the dealership is not having an "out of the door price" first. (Edmunds.com shopper survey 2013-2014) The internet culture from sites like Amazon.com, Target.com, etc... has created shoppers who are used to seeing prices and comparing them. In a recent worldwide survey, **44%** said they are ready to purchase a vehicle on-line if the choice is presented to them. (Capgemini auto shopper survey 2014). By not displaying real time pricing, you are effectively chasing away more than half of your potential customers.

Q. *I do not have the time to update vehicles every day, how is this accomplished?*

A. eznewcar.com is updated every night. Sold vehicles once processed in your dealership are automatically eliminated, and as new inventory arrives, it is incorporated into the flow. Sale pricing is set by vehicle type as either a market up, or down, in dollars or as a percentage of cost.

Q. *How do I know what is happening on my area of the site?*

A. Each dealer has a dashboard available to him. This dashboard shows the dealer personnel a real-time picture of the activity on the site, hits, click-thru's, which cars, which pricing are all shown in daily, and month-to-date displays.

Q. *We have a lower closing ratio of on line shoppers than convention shoppers, why is eznewcar.com better, like you said?*

A. eznewcar.com will have a higher closing ration because shoppers that come to eznewcar.com are ready to buy. The other sites are designed to show vehicles, usually at their MSRP price, get the shoppers contact information, and then send it to a variety of dealers who are then in a "timed" battle to reach that customer with an attractive price. All too often, there are multiple leads generated by the same customer which distort the true number of leads actually generated. Shopper of eznewcar.com usually know the vehicle they want, they are now looking for price to see how it fits into their budget .....they are ready to buy, not shop.

Q. This all sounds fine, but how do you bring shoppers to the site?

A. eznewcar.com is actively marketing, through North Jersey Media, with an aggressive print campaign utilizing their extensive collection of daily, weekly and monthly products to reach hundreds of thousands of shoppers. In addition to that monthly e mail blasts to specific demographics, and sponsorship of monthly automotive story at bergencounty.com and various events throughout the area (North Jersey Test Drive in September at MetLife Stadium). Eznewcar.com has created partnerships in the police, fire and teaching communities to bring access to their hundreds of thousands of employees and Hospital Organizations where their employees shop on line to obtain goods and services. We will always strive to get the most "bang for our advertising buck", in order to keep costs down for the dealers benefit.

Automotive Buyers Study of Trends in Automobile Marketing:

The *2011 Polk Automotive Buyer Study* found that today's car buyers spend more time with their eyes glued to cars on a computer screen than to cars on an actual car lot. According to the study, prospective buyers now spend an average of 11 hours doing research before they set foot on a car lot; and they're only averaging about seven hours at the dealerships. This 24/7 access to information via the internet has drastically tipped the scales in favor of "online real estate" (i.e. computer screens, smart phones, or tablets) over traditional main street USA real estate...

Google's 2011 Automotive Buyer Study found that today's car buyer uses a tablet or smart phone to perform a search over 50 percent of the time. What does this mean in regard to search and your dealership? Well, tablet and smart phone screens average three to nine inches (which is smaller than a normal computer screen by roughly 50 percent), so being found on page one of search engines like *Google, Yahoo,* and *Bing* is increasingly critical; searching on a small screen can be hard on the eyes, especially when trying to scroll though pages and pages of results to find what you are seeking. Only two percent of car buyers actually go past page one of the search results now, so if your ads are not appearing on page one, you are probably getting no exposure to in-market buyers. Using video in search (VSEO) is a great way to be on page one of the search engines—using sight, sound, and motion to connect via prospective car buyers with the most engaging media format available. Just like in the game of Monopoly, he who owns the most property wins…and page one of search is the new "Boardwalk."

One dealership actively using VSEO is Pacific Volkswagen, part of the LAcarGuy family of dealerships in greater Los Angeles. Brad Burlingham, vice president of marketing for the group says "We are using Video SEO to gain many first page Google listings under thousands of keyword combinations for our brands and markets. This enables us to reduce other competing dealer and third-party website listings, and helps us gain an edge to get exposure to prospective buyers." The screenshot with this article for "2012 VW CC TORRANCE" shows that Pacific VW has five videos on the top of the search page, as well as their website listing. This effectively increases Pacific VW's online market share while reducing exposure for their competitors. Also, there is a third-party website listing (outlined in red) that takes prospects to just one web page that has over 21 different dealers listings competing to sell this model. Start using Video SEO today to gain an edge and increase your online market share to in-market buyers.

References taken: AJ LeBlanc, is the cofounder of Car-mercial.com and Carbuyersengine.com. For more information, email aleblanc@dealermark.com or visit www.car-mercial.com and www.carbuyersengine.com.

Eznewcar.com

Preliminary patent outline

A web portal that is designed to display a variety of new and /or used vehicles for sale by car dealers where the user can choose a variety search criteria to find a specific vehicle. The shopper can search by dealership to see all of their offerings, by a specific make and model vehicle from a plurality dealerships offering that particular vehicle and lastly by price. When the shopper chooses by price, he has the ability to select a lease price, a finance, both lease and finance or by the full purchase price. In all cases, the three pricing options are displayed with applicable regional rebates, any down payments, finance interest rates, lease bank fees, lease residuals term of the lease or finance and all dealer fees. This web portal allows shoppers to search for a vehicle in three simple ways, by price, by dealer and by vehicle and model and in all types of search, it brings the shopper to the dealership vehicle display page where all necessary vehicle information, vehicle identification number (VIN), pictures, miles on the odometer if they are required and all optional and standard equipment on this vehicle are displayed, and from this page, a shopper can choose to contact the dealer, not the other way around as all other site require.

A consumer does not need to enter any personal information in order for him to receive the dealer's best price. Other auto buying web site portals require the shopper to input his personal information and then the "lead" is farmed out to the various dealers who pay the lead provider for that information.

Once the dealers receive that information they bombard the shopper with e mails and telephone calls to attempt to earn his business. In eznewcar.com the shopper views the best price deals that the dealer is offering on his entire inventory of vehicles and can select the vehicle, provided with a specific vehicle identification number (VIN) from the listing and see all of the pricing options he has available to him.

Further, once a consumer has settled on a vehicle that he is interested in, he can then by entering a minimum of personal information into eznewcar.com, and not have it go, as of yet to a dealer, have the deal recalculated based upon his credit tier, have a trade value, if a trade is to be part of the deal, determined and added into the deal, net of any payoff information so he will have an accurate price on said vehicle configured to his credit limits. Once he is comfortable with the deal and vehicle he then can have an "intent to purchase" document completed, complete with sales tax, estimated motor vehicle costs and any dealer fees sent to the dealer and printed for his own information. The dealer then and only then contacts the customer to schedule a time that the customer comes to the dealership to finalize and take delivery of the vehicle.

Vehicle data and information is downloaded into eznewcar.com nightly via a third party source to ensure that only the most accurate listing of vehicles is considered for sale. At this point the selling price of the vehicle is calculated based upon the profit or loss margin that the dealer wishes to sell the individual vehicle at. As a vehicle is sold and processed by the selling dealership, it is removed from the feed and is no longer offered for sale. Once the feed is incorporated into the eznewcar.com site, eznewcar modifies said data to conform to the eznewcar.com method of display completely changing that data and making it completely unique and preparatory information that eznewcar.com will display and manipulate as no other automotive site currently does. Once the data and vehicle information is incorporated into the eznewcar.com site it then undergoes an additional modification where the vehicles lease and finance calculations are completed so the vehicle has all current pricing models displayed. In the case of a lease, the most recent residual for that vehicle is used so that the vehicle lease and finance payment options are calculated in the background and all vehicles are updated with the best possible lease and finance payment options for that vehicle daily. This process includes all of the appropriate rebates that apply to that vehicle regionally.

Milestones

Our corporate Product Milestones to date.

- February 10, 2012 -2013 the day the first business model for PRODUCT (EZNEWCAR) was created.
- After months of planning and countless late night meetings the final outline for the programming was completed with initial idea on paper.
- On September 11, 2012 the website domain name for eznewcar.com and eznewtruck.com were procured as product under HP Infosystem office
- After numerous delays, a beta test version was completed in October 2014- 2015
- Our National Sales Director was hired in and relieved of duties
- Preliminary marketing and initial dealer sign-ups were begun.
- Software architecture strategic partnership with Amazon cloud services was reached.
- The site goes live as BETA version
- June 30, 2014, sales were halted until the new system becomes operational which will be September 2015
- **September 13, 2014, title sponsor of the Total Test Drive at MetLife Stadium.**
- 2015 Adoption of a "direct to newspaper" sales approach.
- **Press Release in the "Daily Clips"** launching the sales strategy of direct through Newspapers. Sales presentations begun to New Jersey and surrounding newspapers.
- **2015, Greater Media contract finalized and Signed.**
- Many additional media companies are contacting us for presentations and the Daily Voice is signed on in August 2015 after only one meeting.

How is it going to grow?

In the beginning when I developed the idea of eznewcar.com was the concept of utilizing newspaper sales persons to sell eznewcar.com. My logic was simple; newspaper sales people knew the market and had excellent relationships with all of the dealers in their marketplace. Further, because of the dramatic shift to on-line marketing, away from the traditional print advertising, I felt that a newspaper sales organization would see eznewcar.com as another "arrow in their quiver", another product to sell and supplement their income. Further still, and I think even more important from their perspective is the fact that eznewcar.com is brand new. The entire US was open and available as a fresh sales territory so the sales people had plenty of "fresh meat" to conquest. In the time between early development and just prior to beginning the sales push, my group felt that we would need a National Sales Manager to oversee the sales structure of the company. A gentleman was interviewed and hired, and given authority to develop the sales approach and processes we would need. After a short time it became clear to us that he was not the right person for the job. Clearly a change was necessary, and he was asked to leave last June. Now, I realize this information is not germane to Eznewcar.com's effectiveness and future growth, but I want to be perfectly clear and open with this plan......yes, mistakes were made but they were quickly identified and corrected so as to keep the company moving forward at all times. An additional means of sales of eznewcar.com is through the advertising agencies. Ad agencies do not make any money with any of the other 3rd party sites. Those companies market direct to the dealer and cut the agency out of any opportunity to make a commission and therefore the agency has no incentive help promote the site to the dealer. That is borne out by the LeadiD.com survey which clears indicates that dealers are being told by their "marketing firms" that these sites don't work. We all were not born yesterday, so their motivation is clearly monetary, and since I own an advertising agency I can attest to this fact being very true. Last month we brought in the former Classified Advertising Director of North Jersey Media, a large newspaper group, with over 50 newspapers in NJ and several glossy magazines as a consultant to formulate our "direct to newspaper" sales and marketing campaign. His expertise and relationships have proven to be tremendously important in the development and implementation of a comprehensive sales and organizational structure for selling eznewcar.com to and through newspaper classified sales departments. By sharing the revenue with the newspaper and after training, allowing their sales personnel to market eznewcar.com to their automobile dealers within their market, eznewcar.com can rapidly move into a variety of markets without the added expense of a large sales organization. By utilizing the already established newspaper classified sales departments; eznewcar.com can grow at a much faster paced rate and steadily move into new markets without the excessive expense of a large media campaign. Built into the newspaper contracts is the required newspaper advertising and impression lead campaigns that newspapers currently have available to them. Eznewcar.com needs only train, monitor and support the newspapers and then fill in additional cable and possibly radio advertising campaigns to complement the efforts of the newspapers. In just a matter of a couple of weeks we have presented our initial proposal to 5 newspaper organizations and have second meetings set up with all of them as the proposals have been very warmly received by all. Our plan is to have them replace their on-line auto search portal with a white labeled version of eznewcar.com.

Pricing models for the initial Newspapers

Consider, for a few minutes, our sales and marketing plan and I think you will understand how it is set up for complete success because it is built on an already existing network that has number anxieties as to how they can compete in an increasingly digital world where they were the main-stay with the analog product. Newspapers were the "go to" source for auto buyers in the analog world prior to the on-set of the internet. Along comes the internet and now dealers are finding that the vast majority of auto shoppers utilize the internet for learning about and locating their next auto. The main internet sites have been the auto manufactures and the dealers' sites.

About 14 years ago, the 3rd party sites commonly referred to as "Pure-play" sites, like cars.com and autotrader.com burst onto the scene. But because these were sold directly to the auto dealer, the newspapers and advertising agencies were locked out of that sales and income opportunity. The newspapers had little choice. Continue to lose the print revenue and see if they could recoup some revenue from the sale of internet impressions and/or banner ads on their news website. But none of this revenue even began to right the ship. Something was needed to give the newspapers, and advertising agencies where warranted, the chance to participate in that revenue stream.

Now along comes eznewcar.com. By supplying the most modern and sophisticated web portal in the marketplace today directly to the newspaper, media sites and advertising agencies, we give them the most advanced platform to sell to their dealer network. Once again the newspaper will be at the forefront of auto marketing where they once were. Since the newspaper sales people have a long relationship with the auto dealer community, they can have immediate credibility with the dealer in each marketplace. Eznewcar.com will not have to bring sales people into those markets and will not waste any time learning the dealers and figuring out their needs and requirements. The newspaper sales people will have this information at their finger tips so sales can begin and ramp-up quickly.

The newspaper will have a new product to sell into their marketplace where 100% of the auto dealers will be conquest sales for them. Revenue will grow,

> Depending upon the size of the newspaper and the number of auto dealers in their area, the monthly subscription and payment to eznewcar.com per dealer will vary. On average, small market newspapers will pay around $2,000 monthly and $4,000 for larger markets for monthly subscription plus $300 per dealer signed up.

valuable sales persons will have an opportunity to earn more, making their position at the newspaper more solid with less attrition and newspapers will have the most advance platform in the market.

Once the newspaper pays the monthly license fee and the dealer commission to eznewcar.com per dealer signed, the remainder is their profit. They can grow this with the addition of more dealers and make accurate forecasts for their management.

This clearly is a win –win situation. The newspaper has a unique and advanced product to sell for the first time. They can become one of the auto dealers most valuable allies in the digital world, secure their position and see substantially increased revenues. Eznewcar.com gets a foothold in a number of regional markets quickly, see immediate cash flow. Further, eznewcar.com does not need an expensive sales force and can keep its management to a minimum. Granted, in the beginning eznewcar.com will give the "lions share" of the sale to the newspaper as we need them to get us established in the marketplace, but once we are set and a proven entity, the sales price will rise and our share of the commission will increase so our revenue will continue to increase due to increased sales, increased gross per sale and additional enhancements and products to sell.

Other than the day to day costs of the broadband, cloud storage and security, the only other direct cost to eznewcar.com is between $60 and $100 per dealer depending upon their current status with Homenet.com.

To get an idea of what kind of revenue we are talking about, we will look at what a basic newspaper market can be worth to us. A newspaper with as few as 40 dealers in its territory generates around $170,000 in annual revenue for us. A newspaper with 600 or more dealers (like NJ Advanced Media….The Star Ledger) can generate, at current levels, $2,300,000 for us and $3,400,000 for them annually. Clearly, this is a lucrative venture for all involved. By using this method of sales and marketing, we create partners who need us for their financial stability and we create a substantially larger market with them then we could ever do on our own in a similar timeframe. Built into each contract is their marketing in their marketplace. Further, every vehicle supported on the "white labeled" site for the newspapers is also displayed on the "parent" eznewcar.com site so eznewcar.com grows with every newspaper organization brought into the fold.

Eznewcar.com was created as a strictly new vehicle sales portal, but because the newspapers have dealer clients that want to sell both new and pre-owned vehicles they need a website that can handle both. This is not an issue in any way for us. We currently are suppressing the feed of pre-owned vehicles from HomeNet.com so we are activating that portion of the feed for the newspaper white label sites and continuing to suppress it on the parent site. Now depending upon how much activity the pre-owned site brings, we may activate it for the parent site too, but we will wait and see if that is necessary.

eznewcar.com 5 year Cash Flow Projection

SALES

Media Group	Number of Dealers	Number closed	Quarter 1	Quarter 2	Quarter 3	Quarter 4	Total	Number of Dealers Period 2	Total Revenue	Number of Dealers Period 3	Total Revenue	Number of Dealers Period 4	Total Revenue	Number of Dealers Period 5	Total Revenue
Greater Media	60	42	21,900	29,200	43,800	43,800	138,700								
NJ Herald	29	20	4,045	16,180	24,270	24,270	68,765								
California	60	42	7,300	29,200	43,800	43,800	124,100								
Chicago	400	280		213,000	426,000	426,000	1,065,000								
Advance NJ	400	280		213,000	426,000	426,000	1,065,000								
Texas 1	50	35		29,250	58,500	58,500	146,250								
Texas 2	60	42		34,500	69,000	69,000	172,500								
Lancaster, PA	120	84			23,000	69,000	92,000								
NY Daily News	300	210			107,000	321,000	428,000								
Newsday	150	105			54,500	136,250	190,750								
Texas 3	200	140			108,000	180,000	288,000								
Princeton, NJ	50	35				39,000	39,000								
North Carolina	40	28				24,000	24,000								
Ohio	150	105				81,750	81,750								
West Virginia	120	84				66,000	66,000								
	2189	1532	33,245	564,330	1,383,870	2,008,370	3,989,815	3,232	9,050,440	5,032	14,090,440	7,032	19,690,440	9,532	26,690,440

INVESTMENT

| | | | 500,000 | 500,000 | | | 1,000,000 | | | | | | | | |
| | | | 533,245 | 1,064,330 | 1,383,870 | 2,008,370 | 4,989,815 | | 9,050,440 | | 14,090,440 | | 19,690,440 | | 26,690,440 |

EXPENSES

	Quarter 1	Quarter 2	Quarter 3	Quarter 4	Total	Period 2	Period 3	Period 4	Period 5
Marketing	40,000	350,000	350,000	500,000	1,240,000	1,800,000	2,500,000	3,755,000	4,125,525
Software	300,000	225,000	225,000	250,000	1,000,000	1,400,000	1,500,000	1,800,000	2,500,000
Technical Infrastructure	150,000	150,000	270,000	335,000	905,000	1,000,000	1,250,000	1,750,000	2,450,000
Equipment	6,000	7,500	15,000	15,000	43,500	50,000	65,000	75,000	80,000
Salaries / Benefits	19,500	105,500	234,500	425,750	785,250	1,750,000	2,450,000	3,250,000	5,650,000
Pay Tax		34,815	77,385	140,498	252,698	577,500	808,500	1,072,500	1,864,500
Office / rent	2,400	2,400	5,000	7,500	17,300	30,000	40,000	45,000	55,000
Travel	1,000	7,500	22,500	35,500	66,500	95,000	125,000	150,000	175,000
Misc.		1,000	2,000	2,500	5,500	25,000	35,000	35,000	45,000
Total Expenses	518,900	883,715	1,201,385	1,711,748	4,315,748	6,727,500	8,773,500	11,932,500	16,945,025
Profit / (Loss)	14,345	180,615	182,485	296,623	674,068	2,322,940	5,316,940	7,757,940	9,745,415

Known Elements and Assumptions for eznewcar.com 5 Year Cash Flow

This projection is laid out in quarterly increments for the first year only, years two through five are shown as the annual total only as we cannot anticipate the quarterly sales accurately enough to justify that exercise. In all instances, this projection is based upon the markets as we know them and what our anticipated expenses will be. I have made every attempt to keep these numbers real and accurate.

Sales :   Currently 16 Media Groups have expressed interest in making eznewcar.com their auto search "widget". One has signed on board, four more are in the process of moving this through their corporate structure and all have expressed tremendous interest in this product and how we have structured this product so that they can earn substantial income from it.

Each media company has given us the number of dealerships in their market; this projection uses 70% of the total dealers in any market as active. In the beginning the first few media companies are set at $300.00 per month per rooftop plus a monthly licensing fee of $2,000.00. As the projection advances in future months the per rooftop number increases to $500.00 per and more than likely, as this product becomes more accepted and spread out, the selling price and our share per rooftop will also increase, but that is not factored into the projection.

Expenses: We have broken the expenses down to just the major categories'.

Marketing covers our branded marketing expenses for eznewcar.com, but it is important to remember that our media partners in each location will also be marketing the website and work to drive views to the site, so the numbers displayed here only reflect our expenditures and not those of the media partner, so the overall marketing strategy will be greater.

Software expense is the expected cost to maintain and improve our current software package. The Technical Infrastructure category is for all of the supporting costs: cloud based storage, bandwidth and partner components in our package. Equipment is computers, printers and high-speed access.

Salaries and Benefits are additional sales and support personnel and the hospitalization and as time progresses management will be added. As of now, the stockholders do not draw salary from the company. Payroll taxes are figured at 33% of payroll.

Office and rent is office supplies and the main office rent factor. As more employees are added, additional space will be needed therefore the expense is increasing. Travel is for trade shows and Media Company training as they sign up. And Miscellaneous Expense is just that, and very minor.

As I mentioned before, we are rolling out version 2 as we speak, version 2.1 is coming quickly with the new lease and finance calculations built in to streamline the process. Version 3 will have the personalized lease and finance pricing specific to the customer's credit, version 3.5 will follow shortly thereafter with the full credit application and integration to the many, if not all dealers CRM.

And with version 4, we will introduce eznewtruck.com which will be the premier commercial truck buying site with integration of the up-fitters packages to the truck cab and chassis's so that the businessman will be able to automate this currently time consuming process. Rather than be forced to give up a morning at the truck dealership during their busy work day, they will be able to shop for, compare and up-fit, if necessary their new truck or truck fleet, all from their home or office during their down time.

Eznewcar.com Architecture encourage me to come up with an aggressive plan and schedule to continue to improve and update our products as we move forward.

Build it and they will come……Not! The main reason why we are seeking additional funding is twofold, the first reason is so we can more quickly develop the additional improvements to the software and improve the site, so that we continue to stay ahead of the competition and secondly so we can advertise the product.

Marketing is a key component to business success. Just look at companies like McDonalds or Verizon. You would think their "word-of-month" traffic is substantial, but you can't avoid their marketing. It is everywhere. TV, newspapers, radio, bus tails, billboards, social media, e mail blasts…..you name it….stadiums, train stations, even the placard in your supermarket cart, their marketing is everywhere because they know it is necessary and it works, and it will work for us too. Trucar.com had an IPO and spent on advertising like a drunken sailor…(many years ago I was in the Navy, and I was a drunken sailor for a short time, so I know how they spend) Anyway, eznewcar.com, if you look at the cash flow plans the investment to fulfill these two areas, investment in the software development faster and more traditional advertising to get our message out. Please do not miss understand this, we will spend wisely, after all, I having operated an advertising agency for 12+ years, I know what works and what does not, and I can get better frequency and reach than most as I have the contacts and understanding to ensure we do not advertise haphazardly. Think back, when was the last time you saw an advertisement for Cars.com or Autotrader.com……it has been a while, has it not? I pay attention to these things, and I can't tell you the last time a commercial ran for them. I mentioned before, these companies have become complacent and uninterested in growing their businesses. But eznewcar.com is going to be aggressive. We have a brand new product; we have superior software and a better website. Consumers will love how easy eznewcar.com will be to use, but if they don't know about it, they can't very well use it. We have a TV spot ready to air as soon as we have the funds available to make it worthwhile.

2014 eznewcar.com is the title sponsor for North Jersey Media's Total Test Drive Event at MedLife Stadium (Giants Stadium). We chose this venue because it was cost effective and ties to our product very well. People can come to the event, drive a variety of vehicles. We can have the dealer's inventory on our site and immediately show a shopper the prices on that vehicle, Lease, Finance and Purchase. It is the type of event that fits hand and glove. In fact, as we move into additional markets, I believe that this kind of event could work very well to introduce us to both the dealers and the shopper alike. We get a huge bang for our buck, and that was what I was talking about before, a wise use of our resources so that we always get the most out of our dollar spend.. Our company has that king of culture.

Call it Minimalist. We want to have a small organization that has no duplication, no wasted effort and no additional people when they are not absolutely necessary. By our sourcing our sales department we save thousands every week. Personnel have so many hidden cost attached to them....benefits, sick time, down time, vacations, taxes, the list goes on and on, but by outsourcing the sales function, but carefully monitor it, we get the sales without the hidden costs, one amount, preset, no surprises, less issues and problems that make so much more sense.

Our company philosophy will carry over to marketing....just because it is investment money (someone else's), that is no reason to miss-spend it. We have taken our own time and money and built the beginnings of a large company, we are ready to see it take off now, and for that to happen, we need this serious cash infusion.

Invest to date in eznewcar.com

As I stated, eznewcar.com was created, between then and now eznewcar.com has been built and self-funded by the hard work, personal dedication and self financing of six individuals. 100% of the programming has been paid for with personal funding and sweat equity of approximately $1,500,000 dollars. The man hours dedicated to eznewcar.com group. The design phase, even if a million dollars invested in our state-of-the-art programming, under a different set of circumstances and with a different group of professionals, the cost would have been substantially higher. The search engine in our system is extremely robust, as it needs to be, because once we have a few regions up and running on the system, the demands will be huge. As we open up the all new version of eznewcar.com, we will house over 19,000 new vehicles in the New York, New Jersey metro area alone. As mentioned earlier, our system resides on an Amazon cloud server array. Amazon has contacted us and promised a dedicated team to help ensure that all of our systems and multiple backups work perfectly. They have pointed out that we are the first, who are using the Amazon cloud services to have a system of our size and complexity hosted in the cloud servers. They see the tremendous potential of what we are doing and are watching carefully. As with most aspects of eznewcar.com we are breaking new ground and forging new territory. All marketing pieces, all promotional and sales documentation has been completed by the team. This is a true testament to the dedication and belief in eznewcar.com. Five of these gentlemen have no automotive background, but immediately saw the potential and viability of the project. Without their hard work and dedication this never could have happened. And along the way, they learned valuable lessons about the automobile retail business. The combination of my retail automotive experience and their programming and design abilities we got the most out of our limited numbers.

Himanshu Shah, Partner, Chief Technology Officer

Himanshu is the chief Technical Officer for the product and the person most responsible for the technical superiority of the eznewcar.com platform. His ability to take my vision and thoughts, which are not technical in nature and see how they could be incorporated into this website, is truly amazing. He saw the future as I asked it to be, and gets most of the credit for the simplicity of the operation of eznewcar.com while maintaining the vastly superior complicated underpinnings of the system. He is the most important component in the company when considering the software.

Business and Professional Organizations eznewcar.com has been in the process of cultivation relationships with Unions (Teachers, Fireman, Police, etc) and large organizations like Hospitals and Insurance companies to name 2. Our plan is simple; provide the links to eznewcar.com for their employee benefits website for their employee base. It is our plan to help establish extra discounts for the employees of these companies if they shop through eznewcar.com.

Social Media

As I pointed out above, social media plays an important role in the word of mouth marketing, or grassroots marketing for eznewcar.com. The fact that we connected with Unation should indicate our knowledge of its powerful reach today. Studies have shown that about 12% of social media users pay attention to company run ads on social media, but when the posting is organic and from a friend telling his friend that he had a great experience using something, or buying something, that becomes over 75% acceptance. So the important point of social media is to keep it organic. Build processes that encourage users to share their great experience with eznewcar.com, rather than us run an "Institutional" ad to attempt to convince others to use our site. How you do it, is as important as what you do. Unation is completely and fully adept at doing that and that is why we took the opportunity to partner with them. Also, they will work with us in the Florida market to help make sure we are successful once we move to that market. We will run social media events and promotions in those markets plus because Unation has highly placed connections in the colleges and Universities which are all over the area, we can be involved in a number of college student events that will down the road lead to many sales as these students get out into the business world.

Dealership Back End Functionality

Up to this point I have talked about eznewcar.com and the advantages to the shopper and the dealer alike, but there another feature for the dealer that will help him sell more cars and manage his inventory better. Built-in to the dealer's dashboard in the system is a suite of information that the dealer can use to run his business more effectively. He will have available to him real time data on the number of hits he is getting, what vehicles and what pricing models are working for him. He will better know what vehicles and options he needs to stock and what time of the year to do that. I realize that many ordering restrictions come from the factory, but all too often I have seen in dealerships where an unknowledgeable sales manager orders what he wants, not what the information is telling him to order. We aim to help correct that. Also in a future version, once we have a substantial set of dealers signed on, we will provide brand, or segment analytics for the dealers review. So he can see how he stacks up against the other dealers in his market selling the same product, or even similar products. We plan on becoming more than just a lead provider. Our vision is to be so much more. We want to be completely indispensable to the dealer and the go-to site for the consumer. We plan on capturing as much data as possible so we can have the important statistics real time. I was always astounded in car dealerships. We had very sophisticated computer systems to run the accounting, databases and manufacturer contact. My parts department manager could tell you how many $1.25 spark plugs he sold in a week and the system would automatically reorder them for us…..but the most expensive item for sale in the dealership, the new vehicles, had absolutely no system to help management control and order the vehicles he should. I see that that is being corrected to a point, but eznewcar.com will be able to provide up to date real time information for management to make the most informed decisions possible.

One Last Point in Support of eznewcar.com We own the websites for eznewcar.com and eznewtruck.com, obviously. But we also own a group of additional names that we plan to unitize also. We own eznewford.com, eznewlexus.com, eznewhonda.com, and so on. We own over 27 additional names of all of the brands, including eznewtesla.com, just in case. The plan is as eznewcar.com becomes a mainstay web portal, we can offer to individual manufacturers' a "closed site" just for their brand. So if someone were to go to, say eznewford.com, they would see the offerings of just the Ford dealers without the other brands mixed in. We would manage it for their dealers and of course, all of those dealers would also be on eznewcar.com too.

EZnewcar.com
DEALER'S MANUAL

Welcome to EZnewcar.com

This manual has been created to help you make the most of your experience with **EZnewcar.com.**

Help from **EZnewcar.com** is accessible through the video tutorials or through our support line 1-855-EZNEWCAR (1-855-396-3922) / 7326627525

Visit the **EZnewcar.com** Support Page at *www.eznewcar.com/support* to browse support topics, FAQs or contact a representative.

To schedule an information session with EZnewcar.com representative, please send us an e-mail to: *sales@eznewcar.com / support@hpinfosystem.com*

Appendix to this guide

Getting started with EZnewcar.com..51

Dealer Homepage..53

Account Information...64

Reports..68

Inventory Listing..70

Log-out..71

Getting started

Begin by logging in to your account

EZnewcar.com has separate log-in for different users in one company. Therefore, use credentials, specified under your name in contract.

In order to log-in to your account, please go to www.dealer.eznewcar.com.

You will see the following page:

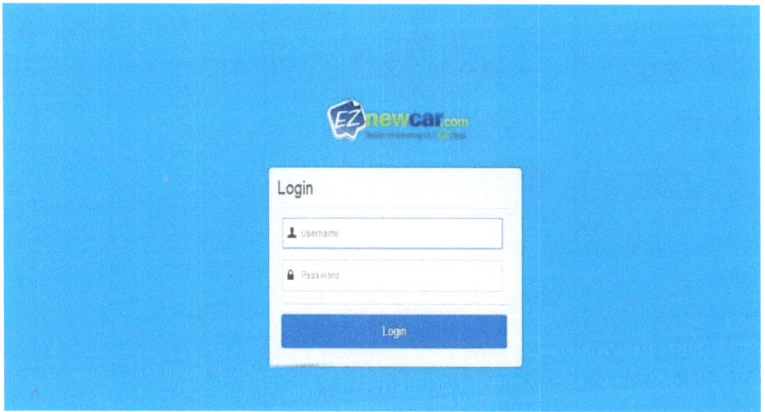

Type in your Username and Password, using credentials from the contract.

In case you do not remember the username or password, please contact us at sales@eznewcar.com / support@hpinfosystem.com or call our support line at: 732-662-7525 / 1-855-EZNEWCAR (1-855-396-3922)

Once, you have successfully logged in, you will see your Account Page.

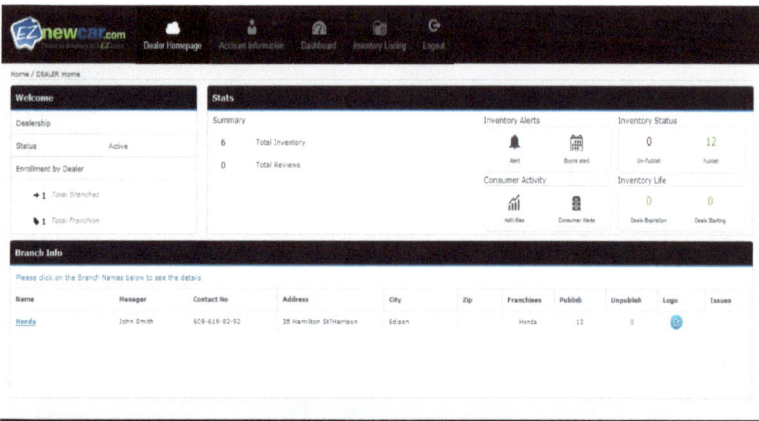

From the icons on top you can choose between:

- Dealer Homepage
- Account Information
- Dashboard
- Inventory Listing
- Logout

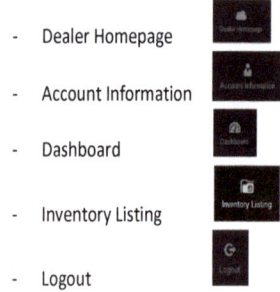

The upper grid contains the following functions:

Homepage icon	Dealer Homepage	Account Information	Reports	Inventory Listing	Logout
When clicking on it, you will be redirected to the Homepage.	The page contain information about all current activities.	Page contains information on your account: security, account summary, contact information	Page contains reports on dealership performance, consumer responsiveness and etc.	This page includes inventory listing for each branch of the dealership.	Click on it to logout from your account.

Let's take a look at each of the pages separately.

Dealer Homepage

Your homepage is your main dashboard, from which you can manage your activity:

- start new promotions
- review alerts
- review branch information

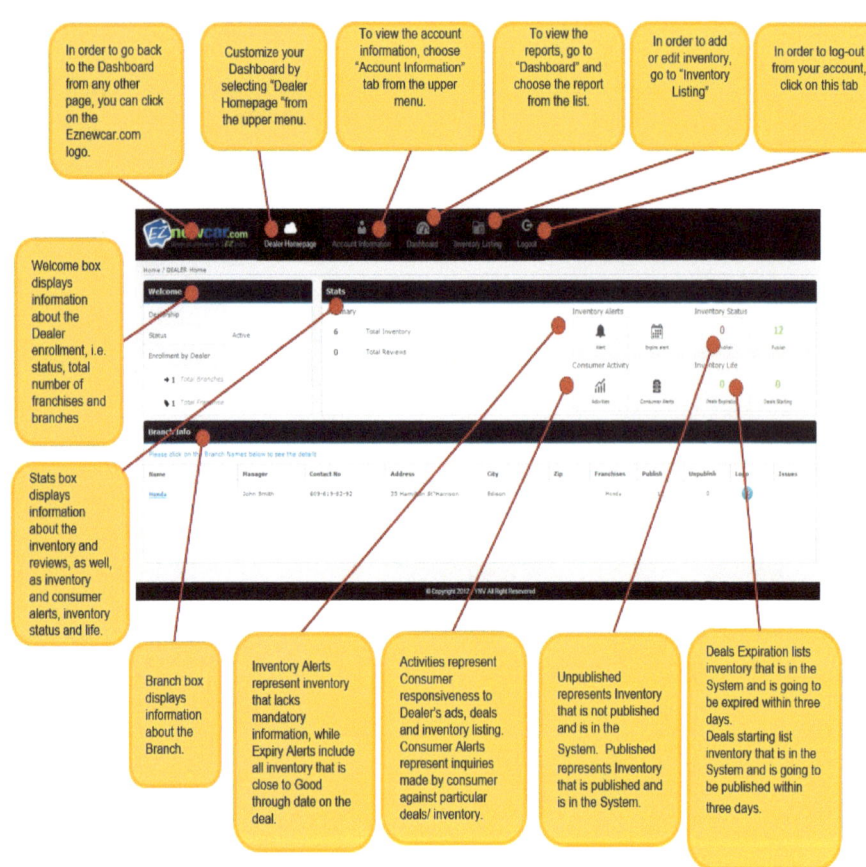

If you want to add a logo to the branch, click on sign under the "Logo" tab and the window will pop out with an "Upload" option:

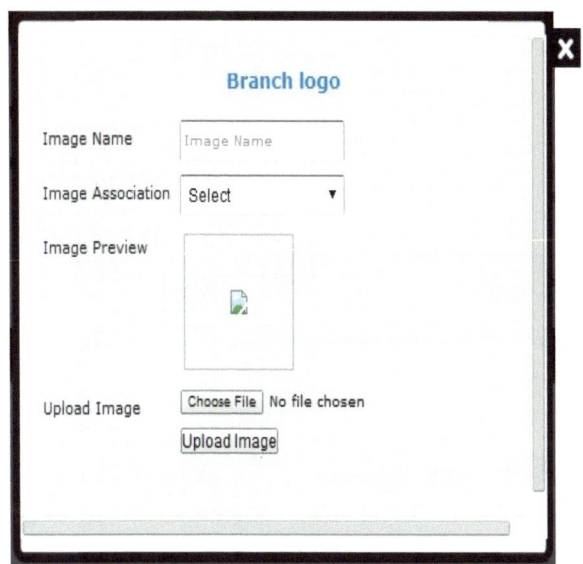

In order for the new logo to be uploaded, you should:

- Type in the Image Name
- Select Image Association from the drop-down menu
- Image Preview allows to view the image the way it is going to be displayed on the page
- Using "Choose File" button, select the file from your computer
- To confirm the selection, click on the button "Upload Image".

To view the Branch details, click on the branch name.

You will be redirected to Branch Details page:

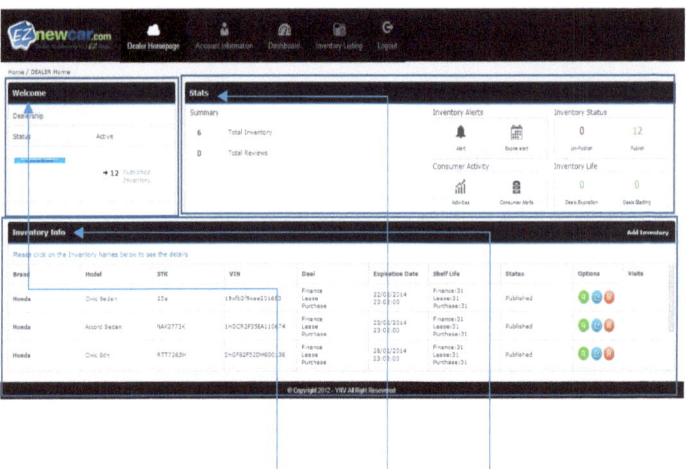

A Branch Details page will include Welcome box, Stats box and Inventory Info box, containing information specific to this Branch.

The Welcome Box has Logo and information about Published Inventory for each particular Branch:

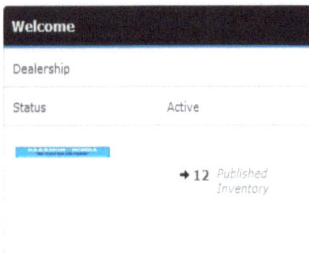

The Stats box on Branch Details page also includes Stats box, which represents:

- Inventory Alerts
- Consumer Activity

#		
1	Total Inventory	Represents the total amount of inventory for a particular Branch.
2	Total Reviews	Represents the amount of review consumers have written for inventories of a particular Branch.
3	Inventory Alert	Lists the inventory that is missing important information and cannot be published until this information is corrected. Can include also inventory that has expired more than 3 weeks ago.
4	Inventory Expire Alert	Lists Inventory that is in the System and is going to be expired within three days.
5	Consumer Activities	Represents data on consumer activities, including reviews, impressions, clicks and etc.
6	Consumer Alerts	Represent messages and alerts by consumers.
7	Unpublished	Provides with the number of inventory that was not published.
8	Published	Provides with the number of inventory that was published.
9	Deals Expiring	Provides with the number of deals that are within three days before expiration.
10	Deals Starting	Provides with the number of inventory that has been added to the system and is starting to be displayed as deals.

- Inventory Status
- Inventory Life

Inventory Info box has inventory listing for a chosen branch with buttons, which allow to view, edit or delete selected information, as well, as to add new inventory.

In order to review inventory entry, click on the ![Q] sign.

View inventory box will pop up:

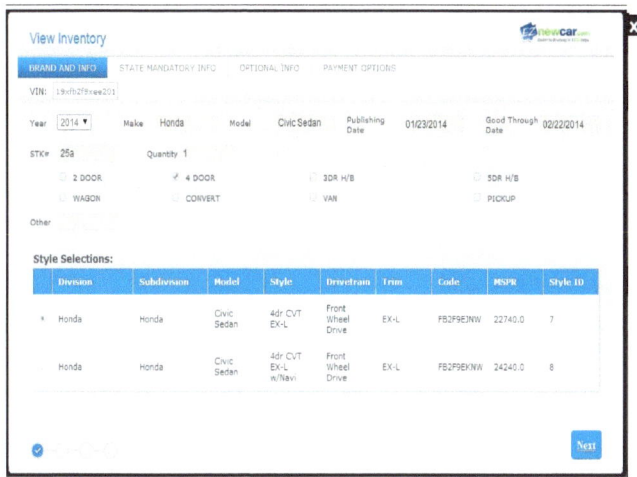

In order to edit it, select ⓔ sign.

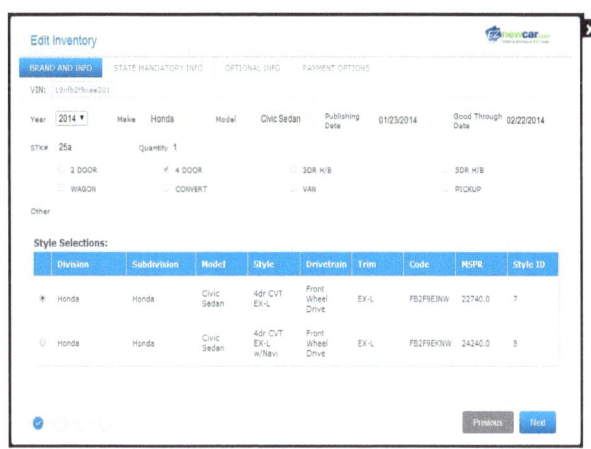

In order to delete an entry, select 🗑 sign.

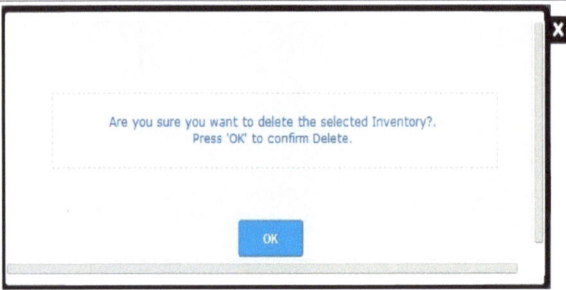

If you would like to ADD INVENTORY to the same Branch, click on "Add Inventory" button in the right corner.

A window will open with spaces for the information to be added for a particular vehicle.

The first page displays BRAND AND INFO.

To start-off, insert 11-digit VIN number in the space provided and click the "Go" button.

System will automatically pull out data, pertaining to a particular vehicle. Publishing Date by default is the beginning of current month. The Good Through Date by default is a month after the Publishing Date. Make sure you adjust them accordingly before proceeding.

In order to go to the next page, fill out the following:

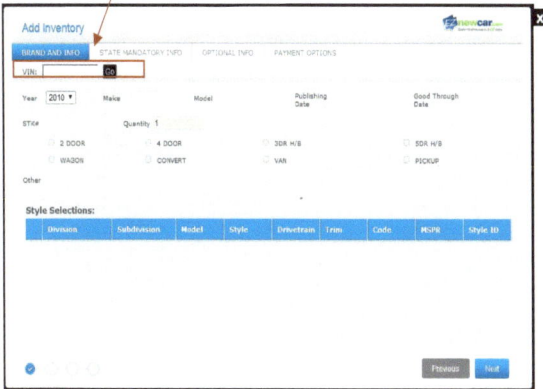

- Stk #
- Publishing Date
- Good Through Date

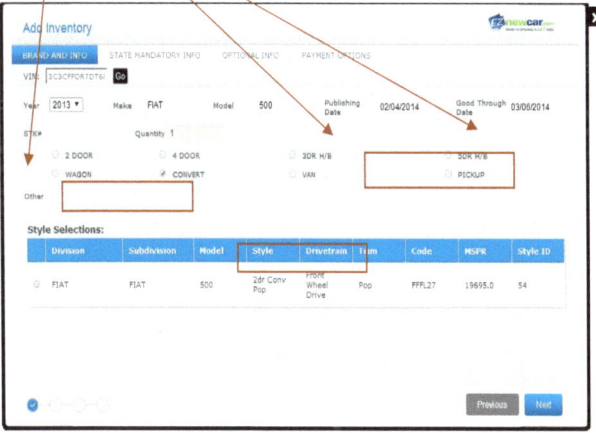

In STATE MANDATORY INFO, System will display information about the vehicle, pertaining to a particular Vin number. You cannot change anything in this section.

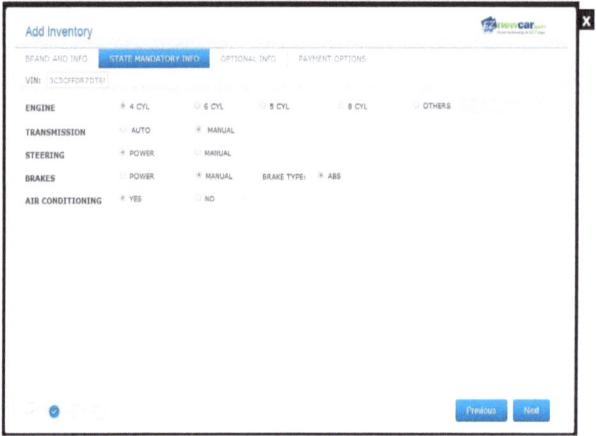

In OPTIMAL INFO, you can check additional boxes, which were not picked by the System.

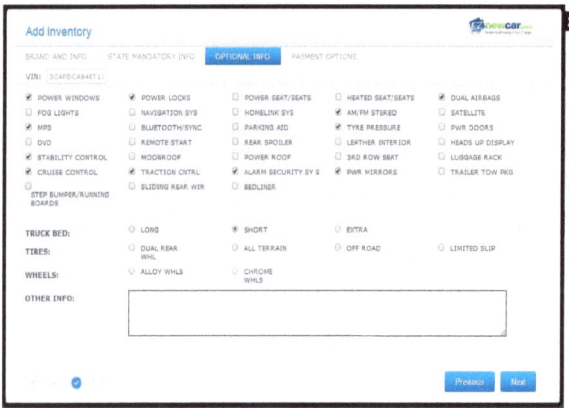

In PAYMENT OPTIONS, you should:

1. Choose between a type of payment/ payments that you want to be displayed for a particular vehicle. You can check all three options.

2. The next step would be to check which option is going to be displayed as the "Hot Deal". The Hot Deal can be only for one pricing option or for two/three.

3. Fill out all the necessary information:

Finance	Lease	Purchase
Monthly Payment	Monthly Payment	Buy for Price
No. of Months	No. of Months	Good Through Date
Down Payment	Down Payment	
APR	Due at Inception	
Balloon Payment	Residual Amount	
Good through Date	Bank Fee	
	Overage Cost	
	Annual Mileage	
	Good through Date	

4. You can add Rebates by clicking on the ➕ sign. You can specify under which Payment Option a certain rebate should be listed and add special comments.

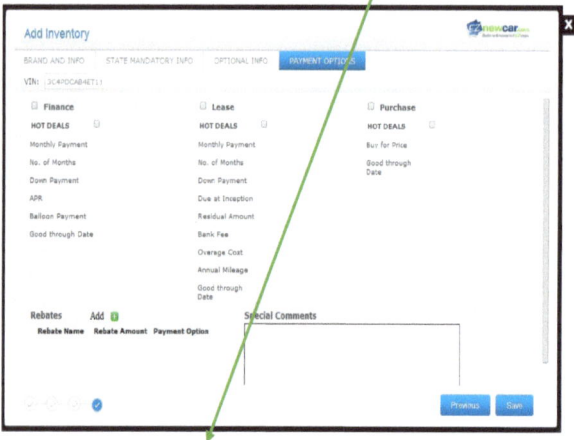

5. After all the information is listed, you can choose to go to the previous page and check all the data before submitting it, click on "Previous" button.

6. When you are ready to submit the data, click on "Save".

Account Information

If you click on "", you will be re-directed to the Account Information page.

The Account Information page looks as the following:

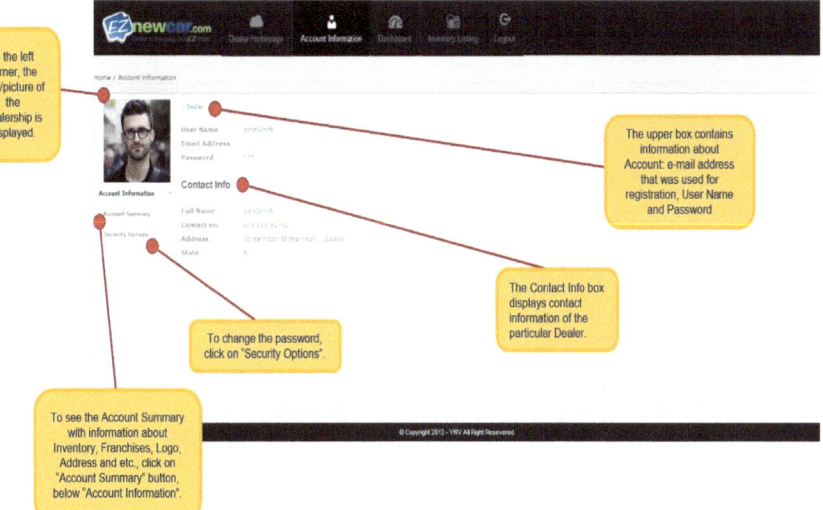

Account Summary

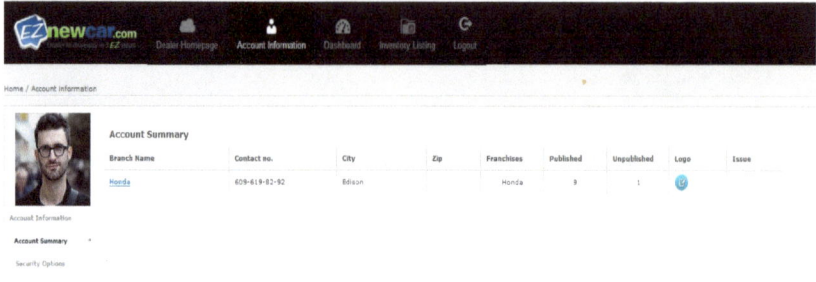

The Account Summary grid contains the following information:

Branch Name	Lists the Branch name
Contact No.	Lists contact number of the Dealer
City	Displays the name of the city
Zip	Displays the zip code of the dealership
Franchises	Lists the name of the franchises the Dealer has
Published	Displays the number of published inventory
Unpublished	Displays the number of unpublished inventory

Logo	Displays the Logo of the Dealership. If you click on ![icon] sign, you can upload a logo.
Issue	Specifies current issues for the particular Dealership.

If you would like to edit any information in this box, please contact sales@eznewcar.com or call our support line 1-855-EZNEWCAR (1-855-396-3922).

If you click on the Branch name, you will get to the Data Summary for the Account page:

Account Summary

1	0 Published Units to date	2	23 Consumer Traffic	3	3 Most Viewed
4	13 Publish Days	5	1 Not Published	6	3 Least Viewed

1	Published Units to date	Displays the amount of Inventory published to date.
2	Consumer Traffic	Displays the amount of people have viewed the inventory, published by the dealer
3	Most Viewed	Displays the number of inventory most viewed
4	Publish Days	Displays the amount of time in days the Inventory has been published on eznewcar.com
5	Not Published	Lists the amount of Inventory that has not been published.
6	Least Viewed	Displays the number of inventory that has been viewed the least.

Security Options

In order to change the password, go to "Security Options". The corresponding page looks as follows:

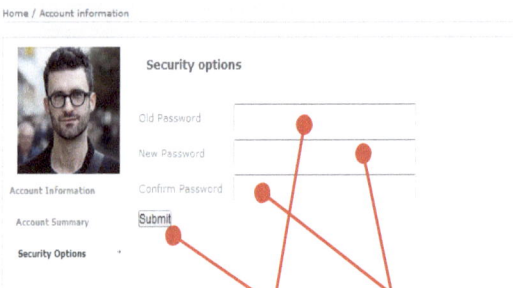

In order to change the password, please follow these steps:

1. Type-in the old password
2. Type-in new password (6-16 characters long)
3. Confirm the new password by typing it in again in the field below.
4. Submit the changes by clicking on corresponding button.

Reports

In order to view the data on Dealership performance, go to "Reports" page, by clicking on "Reports" button: .

A Reports page will open:

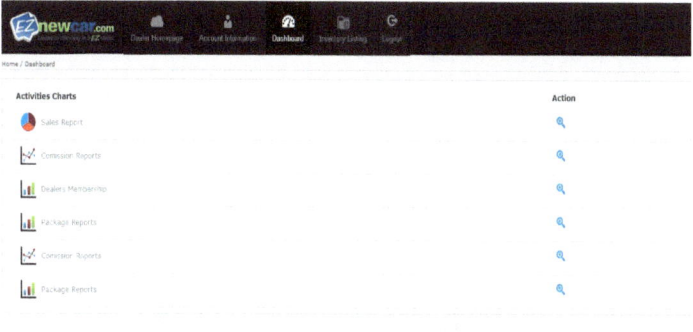

The following reports are present on Reports page:

	Activities Charts	Description
1	Consumer Impressions Report	This report provides with the information on clicks, impressions.
2	Dealer's Performance Report	This report provides with the information on Dealer's performance in terms of consumer preference and deal representation.
3	Inventory Report	This report contains information on inventory: which cars are most/least popular and which ones get the most reviews.
4	Payment Option Report	This report contains graphs and statistics on which payment option is preferred over the other.

In order to view each of the reports, click on the sign next to it: 🔍 under "Action".

Then, a new window will open with appropriate report.

Inventory Listing

If you click on "Inventory Listing" , you are going to see this page:

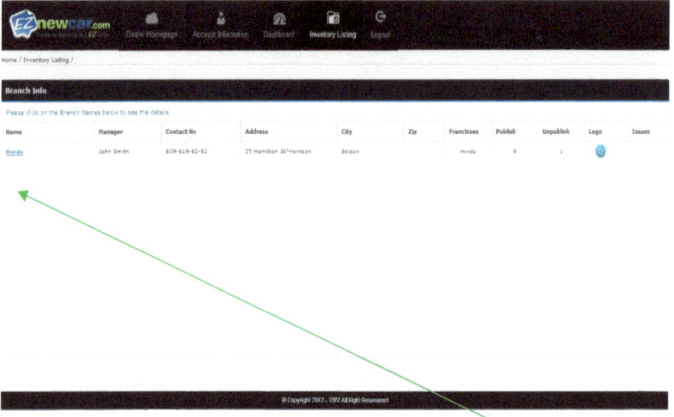

In order to see the Branch details, click on its name. In case several Branches of the Dealership are signed with www.eznewcar.com, choose a Branch name from the list.

At this stage, you can also change the logo of the Branch, by clicking on the sign in the second column from the right.

Branch Name	Lists the Branch name
Contact No.	Lists contact number of the Dealer
City	Displays the name of the city
Zip	Displays the zip code of the dealership

Franchises	Lists the name of the franchises, held by a Dealer
Published	Displays the number of published inventory
Unpublished	Displays the number of unpublished inventory
Logo	Displays the Logo of the Dealership. If you click on ⓔ sign, you can upload a logo.
Issues	Specifies current issues for the particular Dealership.

Log-out

If you wish to log-out from your eznewcar.com account, click on "Log-out" button:

SALES USER MANUAL

Welcome to EZnewcar.com

This manual has been created to help you make the most of your experience with **EZnewcar.com.**

Help from **EZnewcar.com** is accessible through the video tutorials or through our support line 732-662-7525 / 1-855-EZNEWCAR (1-855-396-3922)

Visit the **EZnewcar.com** Support Page at *www.eznewcar.com/support* to browse support topics, FAQs or contact a representative.

To schedule an information session with EZnewcar.com representative, please send us an e-mail to: *sales@eznewcar.com / support@hpinfosystem.com*

Appendix to this guide

Getting started with EZnewcar.com..74

Sales Homepage...76

Account Information..92

Reports...96

Documents...98

Log-out...101

Getting started with EZNEWCAR.COM

Begin by logging in to your account

EZnewcar.com has separate log-in for different users in one company. Therefore, use credentials, specified under your name in contract.

In order to log-in to your account, please go to www.eznewcar.com/sales

You will see the following page:

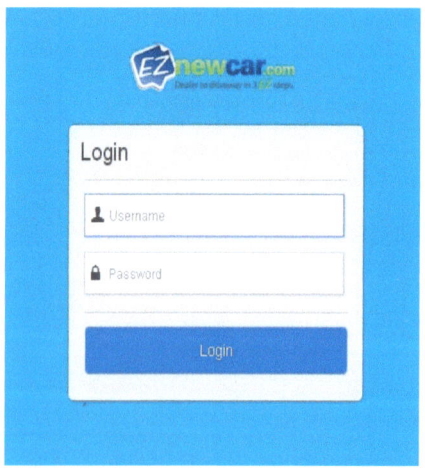

Type in your Username and Password, using credentials from the contract.

In case you do not remember the username or password, please contact us at info@eznewcar.com or call our support line at: 732-662-7525 / 1-855-EZNEWCAR (1-855-396-3922)

Once, you have successfully logged in, you will see your Account Page.

You will see a dashboard menu with the following options:

- Sales Homepage
- Account Information
- Dashboard
- Documents
- Logout

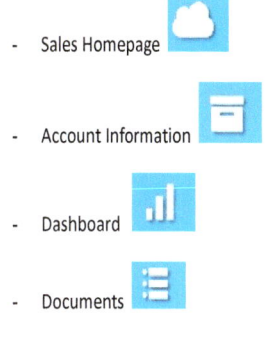

If you click on the "Sales Homepage", you will get to the main Dashboard. From the icons on top you can choose between:

- Sales Homepage
- Account Information
- Dashboard
- Document
- Logout

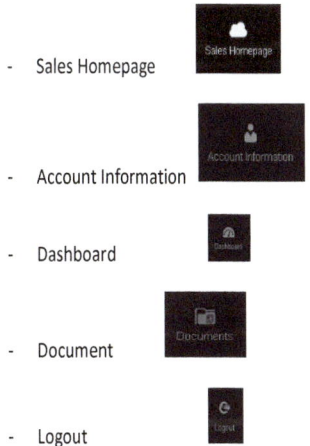

The upper grid contains the following functions:

Homepage icon	Sales Homepage	Account Information	Reports	Inventory Listing	Logout
When clicking on it, you will be redirected to the Homepage.	The page contain information about all current activities for the Sales team.	Page contains information on your account: security, account summary, contact information	Page contains reports on dealership performance, consumer responsiveness and etc.	This page includes inventory listing for each branch of the dealership.	Click on it to logout from your account.

Sales Homepage

Your homepage is your main dashboard, from which you can manage your activity:

- view priority of the dealerships and branches
- view the status of the dealerships and branches
- review branch information
- view the verification status
- enroll new dealerships and branches
- view the latest activities

- view the leads list
- contact dealers and branches
- search
- add dealerships and branches

Sales Home Page

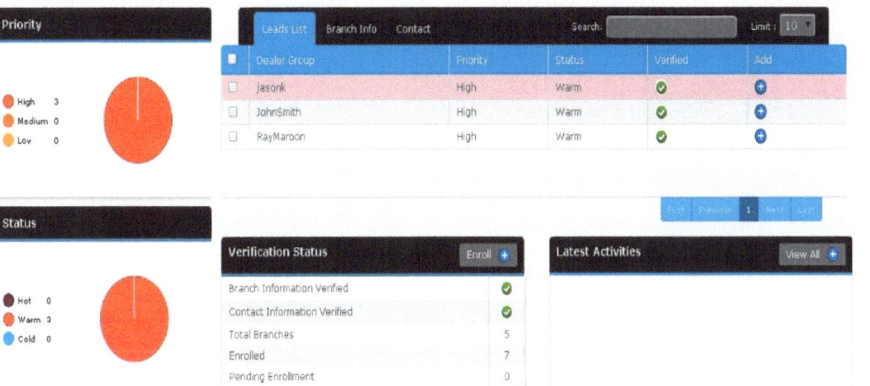

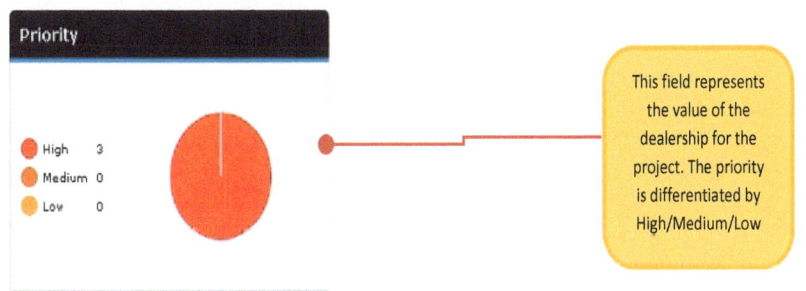

This field represents the value of the dealership for the project. The priority is differentiated by High/Medium/Low

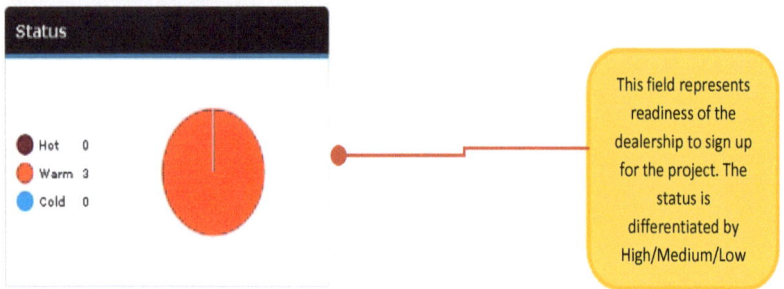

Leads List section highlights:

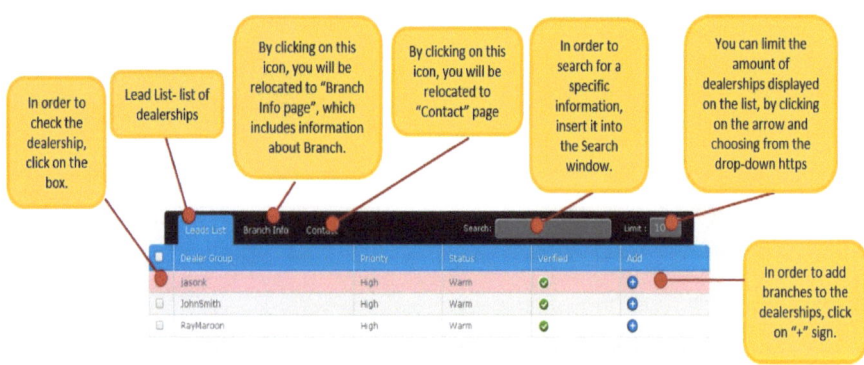

Grid displays the following information:

Dealer Group	The field contains names of dealerships, and branches within, existing in the database, that have been assigned to a particular sales person.
Priority	This field represents the value of the dealership for the project. The priority is differentiated by High/Medium/Low
Status	This field represents readiness of the dealership to sign up for the project. The status is differentiated by High/Medium/Low Hot/Warm/Cold.
Verified	✓ ⊙ The green sign represents that the Dealer is verified. Red that is it unverified. Verified branches and franchises are the ones for which the information has been verified. Verification can also be done by checking the branches and franchises and choosing "Verified" or "Unverified" in the dropdown box. This action usually takes place after the sales person have gone to branches' or franchise's site and verified the contact details.
Add	⊕ By clicking on the sign, you can add a branch, by selecting name, address and franchise information. Only pertains to franchises.

In order to add a Branch, click on the sign "+" in the right column: ⊕ . The following grid will pop up:

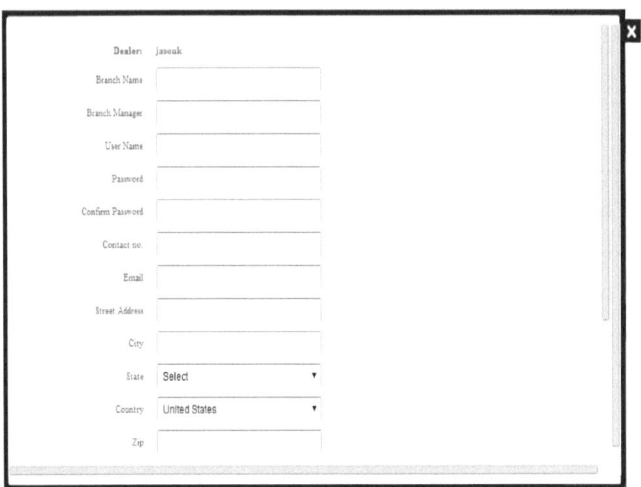

Fill out the grid, using information about the branch.

Dealer	The field displays the name of the dealer, against which you are adding a Branch. Check whether it is the right
Branch Name	Insert the name of the dealership into this field.
Branch Manager	Insert the name of the branch manager into this field.
User Name	Insert the user name for the branch into this field.
Password	Insert the password for the branch into this field.
Confirm Password	Confirm the password by repeating the password you have entered in the previous field.
Contact No.	Insert the Contact No. of the Branch into this field.

Email	Insert the email of the Branch into this field.
Street Address	Insert the email of the Branch into this field.
City	Insert the name of the city where the Branch is located into this field.
State	Select the state from the drop-down list.
Country	Select the name of the country in which the Branch from the drop-down list.
Zip	Insert the zip of the address where Branch into this field.
Franchise Selection	Select the name of franchise from the drop-down list.
Save	Click on "Save" button once you insert all the information.

Once you have saved the information you have entered for a new branch, the Branch will appear under "Branch Info".

Branch Info

If you click on "Branch info", you will get to table will change to Branch Info page, which will display:

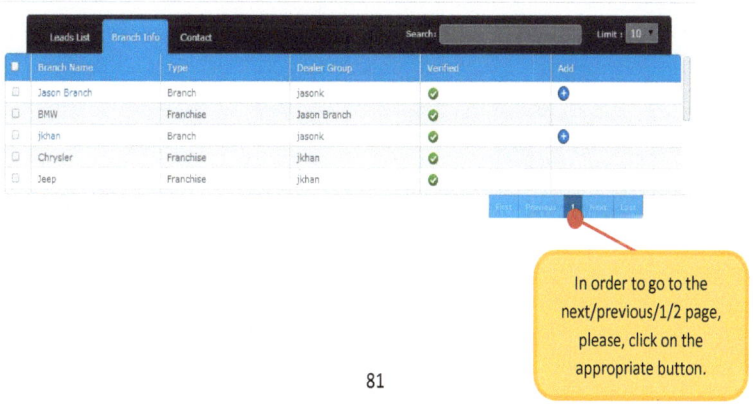

In order to go to the next/previous/1/2 page, please, click on the appropriate button.

Grid displays the following information:

Branch Name	The field represents the name of the Branch.
Type	This field represents the business type: whether it is a Branch, Dealer, or a Franchise.
Dealer Group	This field represents which Dealer Group the Branch belongs to.
Verified	✅ 🔴 The green sign represents that the Branch is verified. Red that is it unverified. Verified branches are the ones for which the information has been verified. Verification can also be done by checking the branches and choosing "Verified" or "Unverified" in the dropdown box. This action usually takes place after the sales person have gone to branches' site and verified the contact details.
Add	➕ By clicking on the sign, you can add a branch, by selecting name, address and franchise information. Only pertains to franchises.

If you click on "Contact", you will get to table will change to Contact page, which will display:

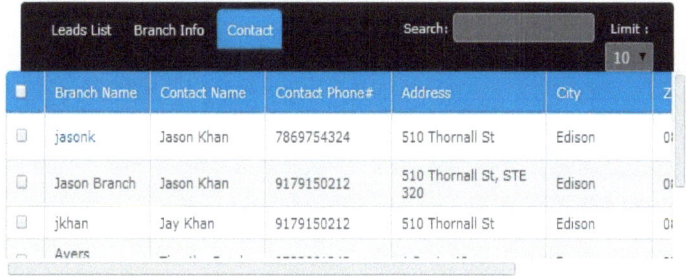

The grid contains the following information:

Branch Name	The field represents the name of the Branch.
Contact Name	This field represents the name of the Contact person in the Branch.
Contact Phone #	The field is required for specifying the Contact phone # for the Branch.
Address	The field is required for specifying address for the branch that is being edited.
City	The field is required for specifying a City for the branch that is being edited.
Zip	The field is required for specifying a Zip for the branch that is being added.
State	The field is required for specifying a State for the branch that is being edited.
Action	This field allows to add a branch or edit information. Edit function is available only for the Dealerships, not for Branches.

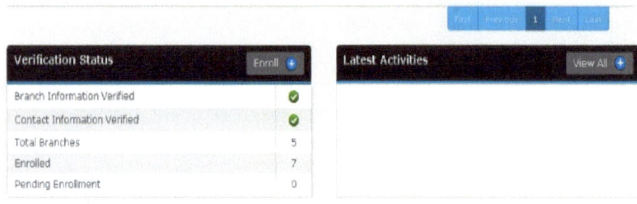

Verification Status

Verification Status	This is the grid that displays the information about verification and allows to enroll a new Dealer or Branch.
Enroll	This field represents the value of the dealership for the project. The priority is differentiated by High/Medium/Low
Branch Information Verified	This field represents whether the Branch information, such as name, address and etc. has been verified by the sales person. If the greed sign is displayed (✓), it means that the information has been verified. If the red sign is displayed (⊘). It means that the information has not been verified.
Contact Information Verified	This field represents whether the Contact information has been verified by the sales person. If the greed sign is displayed (✓), it means that the information has been verified. If the red sign is displayed (⊘). It means that the information has not been verified.
Total Branches	The field displays the total number of branches.
Enrolled	The field displays the total number of branches enrolled.
Pending Enrollment	The field displays the total number of branches that are pending enrollment.

Enrollment

In order to enroll a new Branch, go to Branch Info and check the box next to the Branch you would like to enroll. After that click on "Enroll" button in the grid "Verification Status".

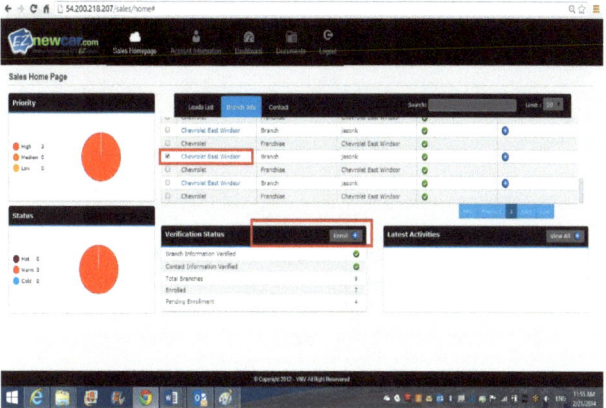

Once you click on "Enroll", a new window will open and you will see the following grid:

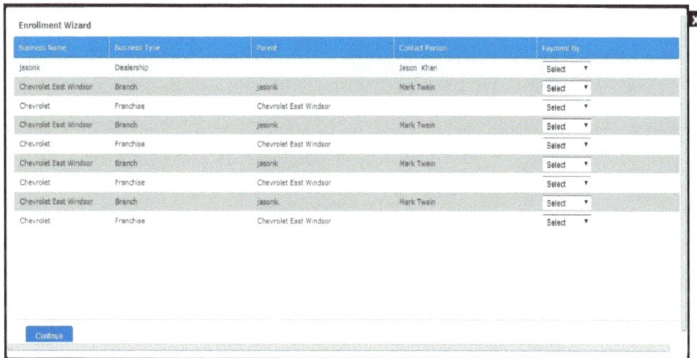

The grid contains the information about the Dealer, Branch and the Franchise. In order to enroll the Branch, you have to specify whether you would like to enroll the dealer will all Branches, or only some Branches of the Dealership, whether the whole Franchise is going to be added or not.

Business name	This grid displays the name of the business.
Business Type	This field displays whether it is a Dealership, a Branch, or a Franchise.
Parent	This field represents the parent entity.
Contact Person	This field displays the name of the contact person of a particular business.
Payment By	The field contains select box with appropriate options to choose from.
Continue	Once you have made selections against each row in the grid, click on "Continue" in order to proceed to the next page.

The payment by options differ by the type of the business:

Dealership	Payment options	Description
Dealership	None / Select / All / **None**	For the Dealership, you can select that the payment is made through it, by selecting "All", or select "None", if the payment is not made by Dealer.
Branch	None / Select / All / **None**	For the Branch, you can select that the payment is made through it, by selecting "All", or select "None", if the payment is not made by Branch.
Franchise	Branch / Select / Dealer / **Branch**	For the Franchise, you can select that the payment is made through the Dealer, or a Branch.

The next page is a "Summary"

The Summary has the following grid:

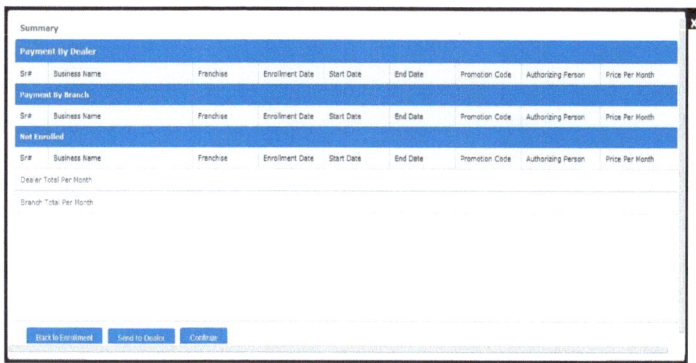

This grid includes the following information:

Payment by Dealer

Sr#	The field provides with the order number of dealers on the list.
Business Name	The field provides with the name of the entity.

Franchise	The field provides with the information, regarding which franchises are included within a particular dealership.
Enrollment Date	Enrollment date is the day the payment for enrollment is processed for the dealer in the project.
Start Date	Start date is the starting date of contract.
End Date	End date is the ending date of a contract.
Promotion Code	The field provides with the code issued by sales supervisor or manager of the project, used for discount in enrollment.
Authorizing Person	This field displays the name of person, responsible for enrollment fee payment.
Price Per Month	The field provides with the price per month specification. This is a price due for the services each month.

Payment by Branch

Sr#	The field provides with the order number of dealers on the list.
Business Name	The field provides with the name of the entity.
Franchise	The field provides with the information, regarding which franchises are included within a particular dealership.
Enrollment Date	Enrollment date is the day the payment for enrollment is processed for the dealer in the project.
Start Date	Start date is the starting date of contract.
End Date	End date is the ending date of a contract.
Promotion Code	The field provides with the code issued by sales supervisor or manager of the project, used for discount in enrollment.
Authorizing Person	This field displays the name of person, responsible for enrollment fee payment.

Price Per Month	The field provides with the price per month specification. This is a price due for the services each month.

Not Enrolled

Sr#	The field provides with the order number of dealers on the list.
Business Name	The field provides with the name of the entity.
Franchise	The field provides with the information, regarding which franchises are included within a particular dealership.
Enrollment Date	Enrollment date is the day the payment for enrollment is processed for the dealer in the project.
Start Date	Start date is the starting date of contract.
End Date	End date is the ending date of a contract.
Promotion Code	The field provides with the code issued by sales supervisor or manager of the project, used for discount in enrollment.
Authorizing Person	This field displays the name of person, responsible for enrollment fee payment.
Price Per Month	The field provides with the price per month specification. This is a price due for the services each month.

Below "Not Enrolled", you can see the following values:

Dealer Total Per Month	The field provides with the total price per month for the Dealer.
Branch Total Per Month	The field provides with the total price per month for the Branch.

After reviewing the Summary, using the bottoms at the bottom of the page, you can either:

- Go back to Enrollment, if you would like to change some information

- Send payment information to the Dealer
- Or Continue to the next page

[Back to Enrollment] [Send to Dealer] [Continue]

Latest Activities

If you click on "View all" (), you will see the following grid:

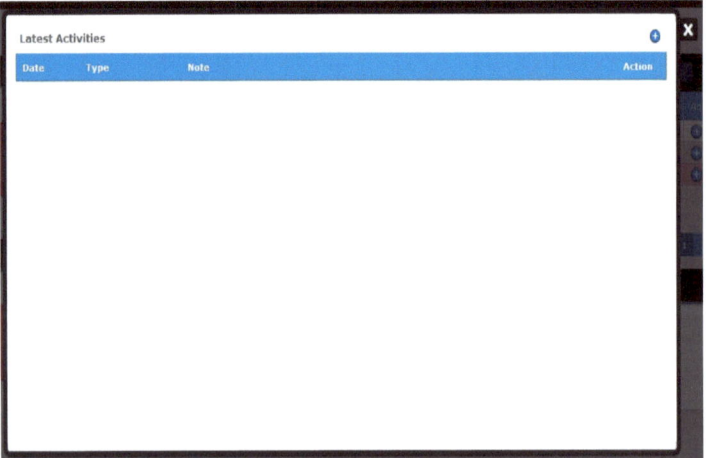

Date	The field represents the date when the note is being added.

Type	This field represents the type of information that is being added (alert/notification/enrollment update)
Note	This field provides with the space for providing details.
Action	➕ ✏️ This field allows to add a branch or edit information in the Note.

Account Information

In order to go to Account Information, please, click on the button "Account Information".

After you click on it, you will be re-directed to Account Information page, which will look as the following one:

Account Summary

If you click on "Account Summary", you will see the following grid:

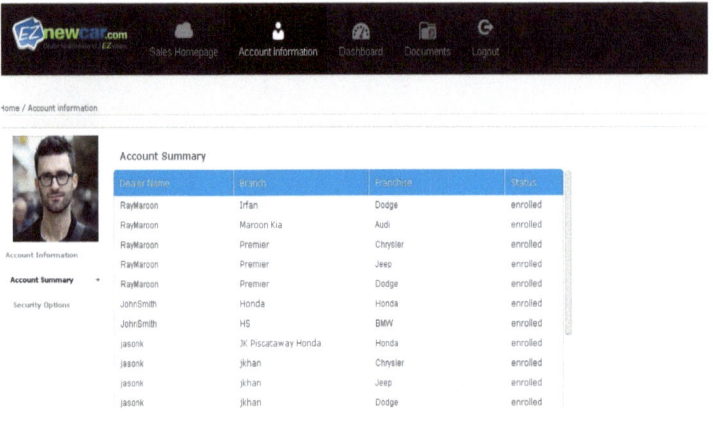

The Account Summary grid contains the following information:

Dealer Name	Lists the Name of the dealer that particular Sales person has in his/her Account
Branch	Lists the Name of the Branch that particular Sales person has in his/her Account
Franchise	Displays the Franchise for a particular Dealership
Status	Displays the enrolment status (enrolled/unenrolled)
Number of Enrolled	In the bottom of the list there is a total number of dealers enrolled by a particular sales person.

| Number of Pending | In the bottom of the list there is a total number of dealers pending enrollment. |

Security Options

In order to change the password, go to "Security Options". The corresponding page looks as follows:

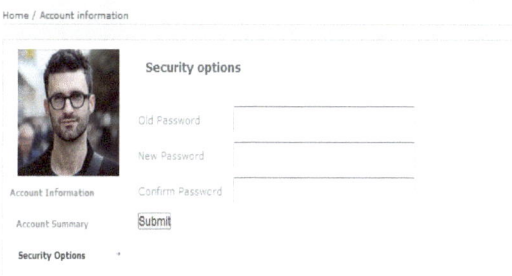

In order to change the password, please follow these steps:

5. Type-in the old password
6. Type-in new password (6-16 characters long)
7. Confirm the new password by typing it in again in the field below.
8. Submit the changes by clicking on corresponding button.

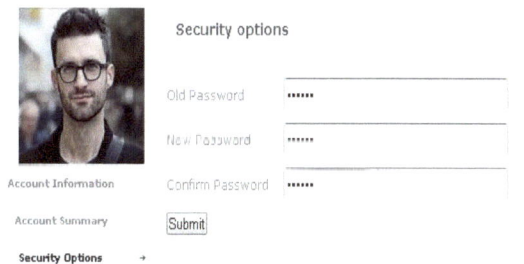

After that, you will have to log-in again, using your new credentials:

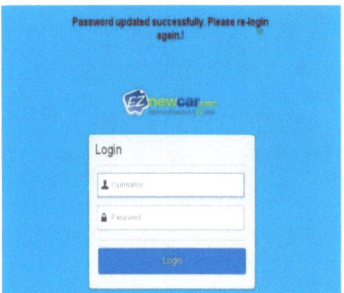

Reports

In order to view the data on Dealership performance, go to "Reports" page, by clicking on "Reports" button: .

A Reports page will open:

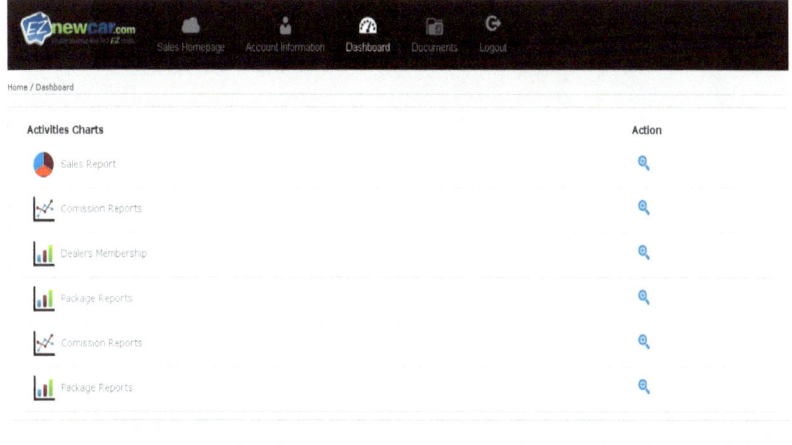

The following reports are present on Reports page:

	Activities Charts	Description
1	Sales per month	This report provides with the information in regards to amount of sales per month.
2	Sales Ranking	This report provides with the ranking of EZnewcar.com sales people.
3	Plan Accomplishment Report	This report contains information on whether a particular sales person has reached the plan or not.
4	Dealer/Branch report	This report contains graphs and statistics on whether this sales person was able to sign up a branch or the whole dealership.

In order to view each of the reports, click on the sign next to it: 🔍 under "Action".

Then, a new window will open with appropriate report.

Documents

In order to review the documents, click on "Documents" button () in the upper menu:

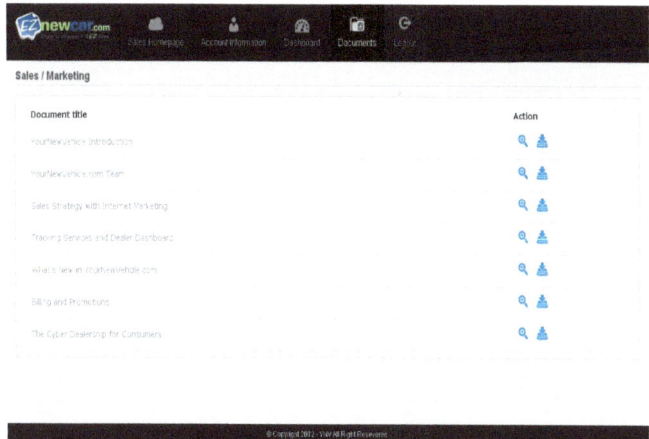

The grid displays the following information:

	Documents	Description
1	EZnewcar.com Introduction	A document provides with introductory information about the product and its history.
2	EZnewcar.com Team	A document introduces the team, involved in production and operation of the system.
3	Sales Strategy with Internet Marketing	This document provides with the Sales Strategy with Internet marketing for sales people to use.
4	Tracking Services and Dealer Dashboard	This document provides with the information on how to use tracking services and dealer dashboard.
5	What's new on Eznewcar.com	This document provides updates on the latest versions of the product.
6	Billing and Promotions	This document provides with the information on Billing and Promotions.

| 7 | The Cyber Dealership for Consumers | This document provides with the information about Cyber online Dealership for Consumers. |

In order to view each of the document, click on the sign next to it: under "Action".

Then, a new window will open with appropriate document. In order to download the document, please, click on " " sign.

Sales / Marketing

Document title	Action
YourNewVehicle Introduction	🔍 📥
YourNewVehicle.com Team	🔍 📥
Sales Strategy with Internet Marketing	🔍 📥
Tracking Services and Dealer Dashboard	🔍 📥
What's New in YourNewVehicle.com	🔍 📥
Billing and Promotions	🔍 📥
The Cyber Dealership for Consumers	🔍 📥

Log-out

If you wish to log-out from your eznewcar.com account, click on "Log-out" button:

The system will display the following window:

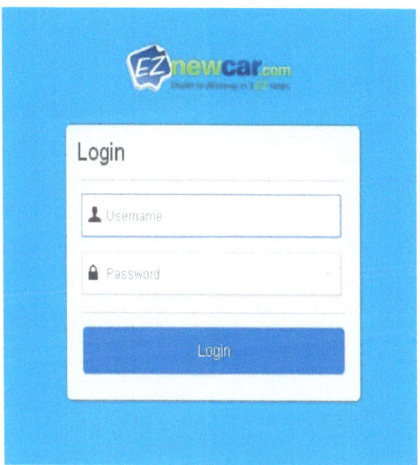

©2014 Eznewcar.com. All rights reserved.

STATEMNENTS IN THIS DOCUMENT REGARDIG THIRD-PARTY STANDARDS OR SOFTWARE ARE BASED ON INFORMATION MADE AVAILABLE BY THIRD PARTIES. INTUIT AND ITS AFFILIATES ARE NOT THE SOURCE OF SUCH INFORMAITON AND HAVE NOT INDEPENDENTLYMADE AVAILABLE AND HAVE NOT INDEPENDENDENTLY VERIFIED SUCH INFOMRATION. THE INFORMATION IN THIS DOCUMENT IS SUBJECT TO CHANGE WITHOUT NOTICE.

Trademarks and Patents

Eznewcar.com, Eznewcar.com logo, and tagline are registered trademarks of Eznewcar.com. Other parties' marks are the property of their respective owners. Features and services within Eznewcar.com products may be the subject matter of pending and issued U.S. patents assigned to Eznewcar.com.

Important

Terms, conditions, features, service offerings and products, specified in this document are subject to change without notice. At Eznewcar.com, we are committed to bringing you great online services through Eznewcar.com. Our services offerings and selection might be occasionally updated, so please check www.eznewcar.com for the latest information, including pricing and availability of our products and services.
SAMPLE TRANDITIONAL NEWS PAPER ADVERTISEMENT IS AS BELOW TAKEN AS REFERENCES ONLY.